SAI SPEED MATH ACADEMY

ABACUS MIND

Step by Step Guide to Excel at Mind Math with Soroban, a Japanese Abacus

LEVEL – 1

INSTRUCTION BOOK

FOR WORKBOOKS 1 AND 2

PUBLISHED BY SAI SPEED MATH ACADEMY

USA

www.abacus-math.com

Copyright ©2014 SAI Speed Math Academy

All rights reserved. No part of this publication may be reproduced, transmitted, scanned, distributed, copied or stored in any form or by any means, electronic, mechanical, photocopying, recording, or otherwise, without prior written permission from SAI Speed Math Academy. Please do not participate in or encourage piracy of copyrighted materials which is in violation of the author's rights. Purchase only authorized editions.

Except in the United States of America, this book is sold subject to the condition that it shall not, by way of trade or otherwise, be lent, re-sold, hired out, or otherwise circulated without SAI Speed Math Academy's prior consent in any form of binding or cover other than that in which it is published and without a similar condition, including this condition being imposed on the subsequent purchaser.

Published in the United States of America by SAI Speed Math Academy, 2014

The Library of Congress has cataloged this book under this catalog number:
Library of Congress Control Number: 2014907005

ISBN of this edition: 978-1-941589-00-7

Thanks to **Abiraaman Amarnath** for his valuable contribution towards the development of this book.

Edited by: WordPlay
www.wordplaynow.com
Front Cover Image: © [Yael Weiss] / Dollar Photo Club
www.abacus-math.com
Printed in the United States of America

Our Heartfelt Thanks to:

Our

Higher Self,

Family,

Teachers,

And Friends

For the support, guidance and confidence they gave us to…

…become one of the rare people who don't know how to quit. (-Robin Sharma)

KIND REQUEST

We believe knowledge is sacred.

We believe that knowledge has to be shared.

We could have monopolized our knowledge by franchising our work and creating wealth for ourselves. However, we choose to publish books so we can reach more parents and teachers who are interested in empowering their children with mind math at a very affordable cost and with the convenience of teaching at home.

Please help us know that we made the right decision by publishing books.

- ❖ We request that you please buy our books first hand to motivate us and show us your support.
- ❖ Please do not buy used books.
- ❖ We kindly ask you to refrain from copying this book in any form.
- ❖ Help us by introducing our books to your family and friends.

We are very grateful and truly believe that we are all connected through these books. We are very grateful to all the parents who have called in or emailed us to show their appreciation and support.

Thank you for trusting us and supporting our work.

With Best Regards,

SAI Speed Math Academy

Dear Parents and Teachers,

Thank you very much for buying this teacher's instruction book. We are honored that you choose to use this instruction book to learn and to teach your child math and mind math using the Japanese abacus called the "Soroban". This is our effort to bring a systematic instruction manual to help introduce children to soroban. This LEVEL – 1 instruction book deals with basics of soroban and with 5 exchange concepts also call small friends.

This book is the product of over six years of intense practice, research, and analysis of soroban. It has been perfected through learning, applying, and teaching the techniques to many students who have progressed and completed all six levels of our course successfully.

We are extremely grateful to all who have been involved in this extensive process and with the development of this book.

We know that with *effort*, *commitment* and *tenacity*, everyone can learn to work on soroban and succeed in mind math.

We wish all of you an enriching experience in learning to work on soroban and enjoying mind math excellence!

We are still learning and enjoying every minute of it!

GOAL AFTER COMPLETION OF LEVEL 1 – WORKBOOKS 1 AND 2

On successful completion of the two workbooks students would be able to:
1. Add any two digit numbers that does not involve carry-over or regrouping problems.
2. Subtract any two digit numbers that does not involve borrowing or regrouping problems.

HELPFUL SKILLS

- Know to read and write numbers 0-99
- Know to identify place value of numbers

PRACTICE WORKBOOKS FOR STUDENTS

There are two workbooks available for students to practice on the concepts given in this LEVEL – 1 Instruction book. Complete Workbook – 1 before proceeding to Workbook – 2. These Workbooks are **sold separately** and are available under the titles:

Abacus Mind Math Level – 1 Workbook 1 of 2 – ISBN: 978-1-941589-01-4
Abacus Mind Math Level – 1 Workbook 2 of 2 – ISBN: 978-1-941589-02-1

WE WOULD LIKE TO HEAR FROM YOU!

Please visit our Facebook page at https://www.facebook.com/AbacusMindMath. Contact us through http://www.abacus-math.com/contactus.php or email us at **info@abacus-math.com**.

We Will Award Your Child a Certificate Upon Course Completion:

Once your child completes the test given at the back of the workbook – 2, please upload pictures of your child with completed test and marks scored on our Facebook page at https://www.facebook.com/AbacusMindMath, and at our email address: http://www.abacus-math.com/contactus.php

Provide us your email and we will email you a personalized certificate for your child. Please include your child's name as you would like for it to appear on the certificate.

LEARNING INSTITUTIONS AND HOME SCHOOLS

If you are from any public, charter or private school, and want to provide the opportunity of learning mind math using soroban to your students, please contact us. This book is a good teaching/learning aid for small groups or for one on one class. Books for larger classrooms are set up as 'Class work books' and 'Homework books'. These books will make the teaching and learning process a smooth, successful and empowering experience for teachers and students. We can work with you to provide the best learning experience for your students.

If you are from a home school group, please contact us if you need any help.

Contents

LEVEL 1 – INSTRUCTION BOOK 1

ATTRIBUTES TO SUCCEED 4

PARTS OF ABACUS 5

STUDENT'S SITTING POSITION 6

CLEARING OR SETTING ABACUS AT ZERO 6

FINGERING 7

ADDING AND SUBTRACTING 8

PLACE VALUE OF RODS 9

ORDER OF OPERATION 9

LESSON 1 – INTRODUCING – EARTH BEADS 10

   LESSON 1 – EXAMPLE 13

   LESSON 1 – SAMPLE PROBLEMS 15

LESSON 2 – INTRODUCING – HEAVEN BEAD 16

   LESSON 2 – EXAMPLE 18

   LESSON 2 – SAMPLE PROBLEMS 20

LESSON 3 – INTRODUCING– NUMBERS 6 AND 7 21

   LESSON 3 – EXAMPLE 22

   LESSON 3 – SAMPLE PROBLEMS 24

LESSON 4 – INTRODUCING – NUMBERS 8 AND 9 25

   LESSON 4 – EXAMPLE 26

   LESSON 4 – SAMPLE PROBLEMS 28

LESSON 5 – INTRODUCING – MIND MATH 29

   MIND MATH – INSTRUCTIONS 30

   LESSON 5 – EXAMPLE 33

   LESSON 5 – SAMPLE – MIND MATH 35

LESSON 6 – MIND MATH – Numbers 6 and 7 .. 36

   LESSON 6 – EXAMPLE 36

   LESSON 6 – SAMPLE PROBLEMS 38

LESSON 7 – MIND MATH – Numbers 8 and 9 .. 39

   LESSON 7 – EXAMPLE 39

   LESSON 7 – SAMPLE PROBLEMS 41

LESSON 8 – SMALL FRIEND COMBINATIONS ... 42

LESSON 9 – INTRODUCING +1 CONCEPT 44

   LESSON 9 – EXAMPLE 44

   LESSON 9 – SAMPLE PROBLEMS 46

LESSON 10 – INTRODUCING –1 CONCEPT 47

   LESSON 10 – EXAMPLE 47

   LESSON 10 – SAMPLE PROBLEMS 49

LESSON 11 – INTRODUCING +2 CONCEPT AND DICTATION 50

   LESSON 11 – EXAMPLE 51

   LESSON 11 – SAMPLE PROBLEMS 53

   WEEK 11 – DICTATION 54

LESSON 12 – INTRODUCING –2 CONCEPT 55

   LESSON 12 – EXAMPLE 55

   LESSON 12 – SAMPLE PROBLEMS 57

LESSON 13 – INTRODUCING +3 CONCEPT 58

LESSON 13 – EXAMPLE 58

LESSON 13 – SAMPLE PROBLEMS 60

LESSON 14 – INTRODUCING –3 CONCEPT 61

LESSON 14 – EXAMPLE 61

LESSON 14 – SAMPLE PROBLEMS 63

LESSON 15 – INTRODUCING +4 CONCEPT 64

LESSON 15 – EXAMPLE 64

LESSON 15 – SAMPLE PROBLEMS 66

LESSON 16 – INTRODUCING –4 CONCEPT 67

LESSON 16 – EXAMPLE 67

LESSON 16 – SAMPLE PROBLEMS 69

LESSON 17 – INTRODUCING – HUNDRED'S PLACE NUMBER 70

USING FORMULAS ON HUNDRED'S ROD 70

LESSON 17 – EXAMPLE 70

LESSON 17 – SAMPLE PROBLEMS 72

ANSWER KEY 74

LEVEL 1 – INSTRUCTION BOOK

TOPICS COVERED

This LEVEL – 1 instruction book deals with the step by step introduction of the basics of soroban and only 5 exchange concepts with examples and sample problems.

Let your students use the Level 1 – Workbook 1 & 2 as you go through the lessons one by one in order given while teaching them.
(Workbook 1 – has more work for the concepts found in Lesson 1 to Lesson 10.
Workbook 2 – has more work for the concepts found in Lesson 11 to Lesson 17.)

INTRODUCTION

Sit at a desk with a comfortable height.
Study the picture and learn the names of the different parts of the abacus.
Practice clearing your abacus a few times.

FINGERING

Correct fingering is very important, so practice moving earth beads and heaven beads using the correct fingers.

INSTRUCTIONS GIVEN WITHIN THE VIOLET BOX

General information to be remembered while working on the abacus is given here.

INSTRUCTIONS GIVEN WITHIN THE RED BOX

Suggests and explains skill building activities that will help improve understanding of that particular week's concept.

INSTRUCTIONS GIVEN WITHIN THE PINK BOX

Useful information in the process of teaching can be found in the pink box.
The most likely mistakes that a child tends to make are also explained here with ways to rectify them as needed.

Each and every child is unique in his/her own respect. His/her understanding of any new concept is also going to be vastly different. So, expect children to surprise you with unique questions and mistakes. With persistent practice, all the hurdles can be overcome.

BEAD COLORS

◇ = Beads that are not involved in the calculation or the game

◆ = Beads that were already in the calculation or the game

◆ = Beads that have just been moved to ADD in the game

◆ = Beads that have just been moved to MINUS from the game

HOW TO EXPLAIN AN EXAMPLE PROBLEM TO STUDENTS

Teachers: Practice the example problems on your abacus.

To Teach: Example 1: +21

1. Call out the number you are going to add or subtract
2. Say what you are doing on the abacus as explained in the 'Action' column of the example. Each example is explained in detail as it should be told and showed to students.

Clear	Problem	Action
Step 1	+ 21	Move **2 earth beads up** to touch the beam on the **tens rod**. Move **1 earth bead up** to touch the beam on the **ones rod**.

HOW TO READ THE HAND PICTURE

While introducing addition and subtraction fact families which total to 5, use your hand to help children understand the relation between numbers (1+4=5, 2+3=5).

While introducing +/- Small Friend Combination formulas (example: +1 = +5 - 4) use your hand to represent heaven bead with 5 fingers as the value of it. While you explain the formula to the student use your hand with fingers out stretched to represent adding of heaven bead and fold fingers when you minus the combination friend.

Example: +1 = +5 - 4

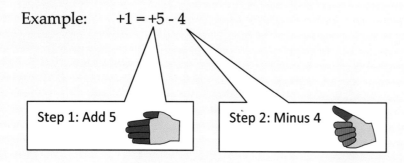

Step 1: Add 5

Step 2: Minus 4

By using fingers it is easy for children to memorize the fact family and also understand the small friend combination formulas. You may choose to use any other items like marbles, counters, beans etc. to teach.

LESSONS 1 to 4 – The goal for the first four weeks is to become familiar with the abacus and understand how a number, when set on abacus, looks. Children need to learn and understand that there are four earth beads and only one heaven bead.

LESSONS 5 to 7 – Mind math (anzan) is introduced. Take time to make sure that the concept is well understood and repeat each problem a couple of times to help with building skills.

LESSON 8 – Small friend combination facts are introduced.

LESSON 9 and 10 – Small friend's formulas for +1 and -1 are introduced. Study the examples given with the descriptions. Work through the sample problems a few times before writing down the answer.

LESSON 11 – Dictation introduced

LESSONS 11 to 16 – Small friend's formulas for +2, +3, +4, -2, -3, and -4 are introduced. Study the examples given with the descriptions. Work through the sample problems a few times before writing down the answer.

LESSON 17 – Using of small friend formula on hundred's place value rod introduced.

The small friends concepts introduced in this level are like foundation for a building, thorough understanding of these facts is vital for their success with 10 exchange concepts, multiplication and division taught in higher levels. So, please make sure that students understand LEVEL – 1 concept very well.

ATTRIBUTES TO SUCCEED

ATTRIBUTE	DESCRIPTION	PARENTS/TEACHERS	CHILDREN
INTEREST	The state of wanting to know or learn about something or someone	Teaching to instill natural curiosity in children.	Learning to look to their adult teachers for guidance about the things they are curious about.
COMMITMENT	Pledge or bind (a person or an organization) to a certain course or policy	• Setting a time and place for your children every day to practice on their abacus without distractions. • Reading the instructions, and guiding and teaching until the student understands the concept.	• Learning that they are acquiring unique skills that many in their peer group do not possess. • Enthused when they understand that they are able to work out their math problems much faster without the use of calculator.
PATIENCE	The capacity to accept or tolerate delay, trouble, or suffering without getting angry or upset.	Directing children's attention back to work when they get sidetracked by TV or other distractions.	Slowly developing with consistent practice.
TENACITY	Persistent determination	"Winners never quit, quitters never win." • Teaching the value of consistent hard work with positive encouragement.	Developing the habits of persistence and hard work.
GUIDANCE	Supervised care or assistance	Providing instruction and guidance until the student learns to work independently.	Reading and understanding concepts, and ultimately working independently
REWARD	A thing given in recognition of service, effort, or achievement.	• Developing and strengthening the parent child bond by spending time together. • Teaching and enriching your child's life and learning experience.	• Learning a lifelong skill • Improving concentration • Enhancing memory power • Gaining self confidence • Developing positive self esteem

PARTS OF ABACUS

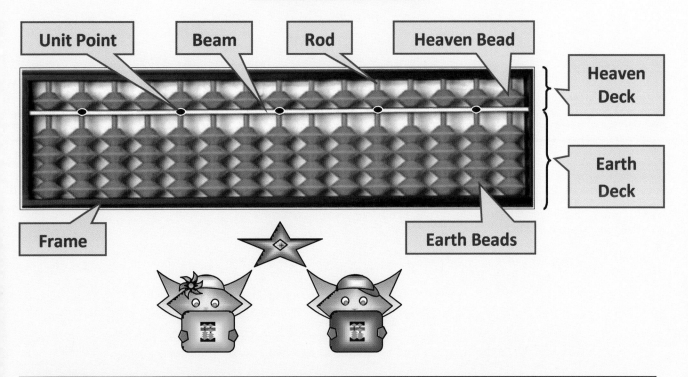

FRAME	Holds all the rods, beads and beam in place.
ROD	Sticks that hold the beads. Beads slide up or down on the rod.
BEADS	Represents numbers on the abacus. They slide on the rod and touch the beam or the frame. When beads touch the frame then your abacus is set at zero.
BEAM	The bar (usually white) that runs across all the rods and separates the Heaven and Earth beads. Only when beads touch the beam do they have value.
UNIT POINT	Can be used as decimal point. Can be used as comma that separates numbers by thousands. Example: $ 102,387,555 = One hundred and two million, three hundred eighty seven thousand, five hundred and fifty five dollars.
HEAVEN BEAD	There is one heaven bead above the beam on each rod. Each heaven bead is equal to "five".
EARTH BEADS	There are four earth beads below the beam on each rod. Each earth bead is equal to "one".

SAI Speed Math Academy

STUDENT'S SITTING POSITION

Sit at a table and place the abacus on the table with the heaven bead deck away from you. Please make sure that the table is not too high for the student.

CLEARING OR SETTING ABACUS AT ZERO

TRADITIONAL METHOD

STEP 1: Place the abacus on the table.

STEP 2: Lift it with the bottom frame still touching the table. This will send all the earth beads to "zero" position.

STEP 3: Gently place the abacus back on the table without moving the earth beads.

STEP 4: Then place your finger between the heaven bead and the beam near the left hand side of the abacus.

STEP 5: Drag your finger along the beam till to you reach the other side of the frame.

This will clear the heaven bead and send them to "zero" position

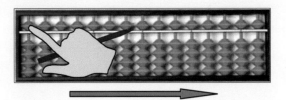

FUN METHOD: ZOOM AND CLEAR

STEP 1: Hold your thumb and pointer finger touching each other.

STEP 2: Place your fingers on the right side of the abacus beam with the beam in between the fingers like you are holding them very gently.

STEP 3: Now hold your abacus with the left hand so that the abacus does not move.

STEP 4: Now gently glide your fingers while still holding the beam, from the right side of the frame to the left side of the frame.

Clear your abacus every time you start a new calculation.

FINGERING

JOB OF THE THUMB (1):

A. Used to push the Earth Beads up to the beam, adding them to the game (ADD).

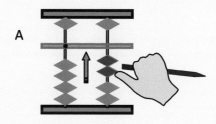

JOB OF THE POINTER FINGER (3):

1. Used to push the Earth Beads away from the beam, removing them from the game (MINUS).
2. Used to push the Heaven bead down to touch the beam, adding it to the game (ADD).
3. Used to push the Heaven bead away from the beam, removing it from the game (MINUS).

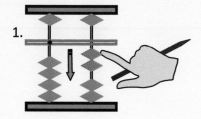 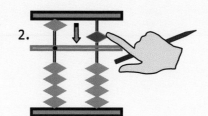 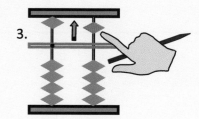

JOB OF THE OTHER THREE FINGERS:

Use your other three fingers to hold your pencil with the point facing down or away from you.

SAI Speed Math Academy

ADDING AND SUBTRACTING

ACTION FOR	ADDING	MINUS OR SUBTRACTING
EARTH BEADS	<ul><li>When we say "adding" then it means we are moving the earth bead up to the beam with our thumb.</li><li>When the earth bead is touching the beam then it is "in the game" and is included in the reading.</li></ul>	<ul><li>When we say "minus" then it means we are moving the earth bead away from the beam and making it touch the frame or other beads that are not in the game with your pointer finger.</li><li>When the earth bead touches the frame then it means that the bead is "out of the game" and is not read.</li></ul>
HEAVEN BEADS	<ul><li>When we say "adding" it means we are moving the heaven bead with pointer finger to touch the beam.</li><li>When the heaven bead touches the beam it means that it is in the game and is included in the reading.</li></ul>	<ul><li>When we say "minus" then it means we are moving the heaven bead with our pointer finger away from the beam to make it touch the frame.</li><li>When the heaven bead touches the frame then it means that the bead is "out of the game" and is not read.</li></ul>

PLACE VALUE OF RODS

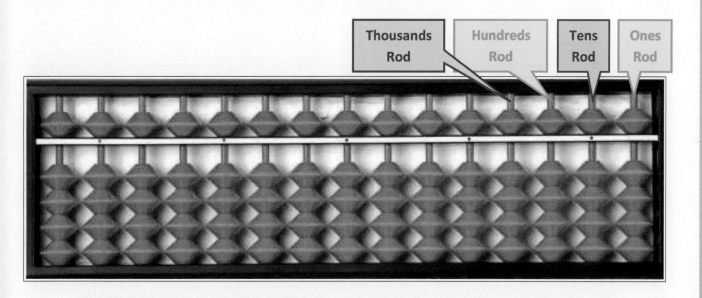

First rod from right is for the **ones place number**. All the ones place number will be set on this rod.

Second rod from right is for the **tens place number**. All the tens place number will be set on this rod.

ORDER OF OPERATION

LEFT TO RIGHT: When working with two digit numbers: always add or subtract the tens place number first and then work on the ones place number.

ATTENTION

The rod near the second unit point can be assumed as a ones rod, too. Sometimes younger students are confused with it if the ones place number is a zero. They sometimes forget to read it. But, with practice, that mistake can be overcome. To make it easy on children, we will assume the first rod on the right to be our ones place rod.

LESSON 1 – INTRODUCING – EARTH BEADS

1) Value of each earth bead = 1

2) There are four earth beads on each rod.

3) Use your thumb to move the earth bead up (adding) to touch the beam.

4) Use your pointer finger (index finger) to move the earth bead down (subtracting) to touch the frame.

5) Always set abacus to zero by clearing all the beads away from the beam before starting each calculation.

6) Setting numbers on the abacus:

   Tens place numbers go on the tens rod.

   Ones place numbers go on the ones rod.

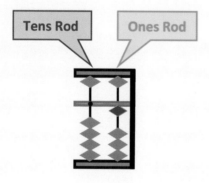

LESSON – 1

NUMBERS ON ABACUS USING ONLY EARTH BEADS

Setting ONES PLACE NUMBERS on the ONES PLACE ROD of the abacus

1 = Push one earth bead on the ones rod up to the beam.

2 = 1 + 1 Push two earth beads on the ones rod up to the beam.

3 = 1 + 1 + 1 Push three earth beads on the ones rod up to the beam.

4 = 1 + 1 + 1 + 1 Push four earth beads on the ones rod up to the beam.

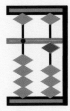   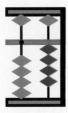

Setting TENS PLACE NUMBERS on the TENS PLACE ROD of the abacus

10 = Push one earth bead on the tens rod up to the beam.

20 = 10 + 10 Push two earth beads on the tens rod up to the beam.

30 = 10 + 10 + 10 Push three earth beads on the tens rod up to the beam.

40 = 10 + 10 + 10 + 10 Push four earth beads on the tens rod up to the beam.

 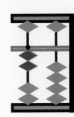 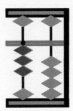 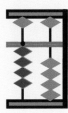

Practice setting and reading the above numbers 1, 2, 3, 4, 10, 20, 30, 40. After your child is comfortable setting and reading these numbers proceed to introduce them to the other numbers found in the list given on the next page.

SAI Speed Math Academy

More numbers that need only earth beads.

11	12	13	14
21	22	23	24
31	32	33	34
41	42	43	44

PRACTICE SETTING NUMBERS WITH EARTH BEADS

Please do the following two activities with your child every day to help him/her become familiar with setting and reading numbers on the abacus.

Use the thumb to add or set the number on the abacus.

00	01	02	03	04
10	11	12	13	14
20	21	22	23	24
30	31	32	33	34
40	41	42	43	44

Activity 1: On the abacus, set any number from the above list and ask your child to read it.

Activity 2: Call out any number from the above list and ask your child to set it on the abacus.

(Do these practices with any 10 numbers from the above list.)

LESSON 1 – EXAMPLE

POINTS TO REMEMBER

Value of each earth bead = 1

Adding earth bead: Use thumb to push the earth bead up to touch the beam or other beads that are already touching the beam.

Subtract earth bead: Use pointer finger (index finger) to push the earth bead away from the beam to touch the frame or other beads that are already touching the frame.

Zero: Always set abacus to zero by clearing all the beads away from the beam before starting each calculation.

Order of action: Set tens place number first and then set ones place number on the abacus.

EXAMPLE: 1

1	After	00	+ 21	+ 01	+ 11	+ 01	= 34
21							
01	ABACUS						
11	LOOKS						
01	LIKE						
34							

Clear	Problem	Action
Step 1	+ 21	Move **2 earth beads up** to touch the beam on the **tens rod**. Move **1 earth bead up** to touch the beam on the **ones rod**.
Step 2	+ 01	There is nothing to do on the tens rod because tens place number is zero. Move **1 earth bead up** to touch the bead which we added in the last step on the **ones rod**.
Step 3	+ 11	Move **1 earth bead up** to touch the bead which we added in the last step on the **tens rod**. Move **1 earth bead up** to touch the bead which we added in the last step on the **ones rod**.
Step 4	+ 01	There is nothing to do on the tens rod because tens place number is zero. Move **1 earth bead up** to touch the bead which we added in the last step on the **ones rod**.

EXAMPLE: 2

1	After	00	+ 34	- 02	- 12	- 10	= 10
34							
- 02	ABACUS						
- 12	LOOKS						
- 10	LIKE						
10							

Clear	Problem	Action
Step 1	+ 34	Move **3 earth beads up** to touch the beam on the **tens rod**. Move **4 earth beads up** to touch the beam on the **ones rod**.
Step 2	- 02	There is nothing to do on the tens rod because the tens place number is zero. Move **2 earth beads down** to touch the frame on the **ones rod**.
Step 3	- 12	Move **1 earth bead down** to touch the bead which is touching the frame on the **tens rod**. Move **2 earth beads down** to touch the bead which is touching the frame on the **ones rod**.
Step 4	- 10	Move **1 earth bead down** to touch the bead which is touching the frame on the **tens rod**. There is nothing to do on the ones rod because the ones place number is zero.

EXAMPLE: 3

1	After	00	+11	+23	- 14	+03	= 23
11							
23	ABACUS						
- 14	LOOKS						
03	LIKE						
23							

LESSON 1 – SAMPLE PROBLEMS

1	2	3	4	5	6	7	8
02	30	11	23	32	33	22	34
01	10	22	- 21	- 12	- 20	12	- 11
01	- 20	01	32	22	11	10	- 11
			- 23	- 12	- 20	- 42	- 11
4	20	32		30		2	1

1:1

1	2	3	4	5	6	7	8
11	33	43	42	33	11	33	43
31	01	- 11	01	- 13	21	10	- 12
- 12	- 04	- 20	- 33	24	02	- 23	01
11	12	01	24	- 44	- 23	14	12
41	42	13	34	02	12	34	44

1:2

1	2	3	4	5	6	7	8
12	22	32	31	41	44	12	14
11	- 11	- 10	02	01	- 21	11	- 02
- 13	22	- 10	- 22	02	- 11	11	21
12	- 11	20	01	- 21	- 02	- 33	- 30
22	22	32	42	23	10	10	3

1:3

ATTENTION

- Some students tend to work with the ones place number first and then the tens place number. Always add or subtract the tens place number first and then work on the ones place number.

- When there is no symbol in front of a number, then it means that the number is to be added.

- In the beginning children will try to **show** you the number instead of adding or subtracting a given number.

LESSON 2 – INTRODUCING – HEAVEN BEAD

1) Value of heaven bead = 5

2) There is one heaven bead above the beam (heaven deck) on each rod.

3) Use your pointer finger (index finger) to move the heaven bead down (adding) to touch the beam.

4) Use your pointer finger (index finger) to move the heaven bead up (subtracting) to touch the frame.

5) Always set abacus to zero by clearing all the beads away from the beam before starting each calculation.

6) Setting numbers on the abacus:

   Tens place numbers go on the tens rod.

   Ones place numbers go on the ones rod.

LESSON – 2

NUMBERS ON ABACUS USING HEAVEN BEAD

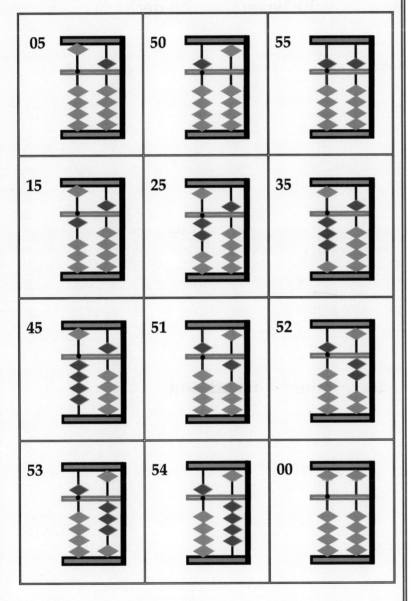

PRACTICE SETTING NUMBERS

Please do the following two activities with your child every day to help him/her become familiar with setting and reading numbers on the abacus.

Use the pointer finger (index finger) to add or subtract the heaven bead on the abacus.

00	10	20	30	40	50
01	11	21	31	41	51
02	12	22	32	42	52
03	13	23	33	43	53
04	14	24	34	44	54
05	15	25	35	45	55

Activity 1: On the abacus, set any number from the above list and ask your child to read it.

Activity 2: Call out any number from the above list and ask your child to set it on the abacus.

(Do these practices with any 10 numbers from the above list.)

SAI Speed Math Academy

LESSON 2 – EXAMPLE

POINTS TO REMEMBER

Value of heaven bead = 5 (imagine that there are 5 beads hiding inside one heaven bead)

Adding heaven bead: Use your pointer finger (index finger) to push the heaven bead down to touch the beam.

Subtract heaven bead: Use your pointer finger (index finger) to push the heaven bead away from the beam to touch the frame.

Zero: Always set abacus to zero by clearing all the beads away from the beam before starting each calculation.

Order of action: Set tens place number first and then set ones place number on the abacus.

EXAMPLE: 1

1	After	+ 25	+ 10	- 30	+ 50	= 55
25						
10	ABACUS					
- 30	LOOKS					
50	LIKE					
55						

Clear	Problem	Action
Step 1	+ 25	Move **2 earth beads up** to touch the beam on the **tens rod.** Move the **heaven bead down** to touch the beam on the **ones rod.**
Step 2	+ 10	Move **1 earth bead up** to touch the bead which we added in the last step on the **tens rod.** There is nothing to do on the ones rod because the ones place number is zero.
Step 3	- 30	Move **3 earth beads down** to touch the bead that is touching the frame on the **tens rod.** There is nothing to do on the ones rod because the ones place number is zero.
Step 4	+ 50	Move the **heaven bead down** to touch the beam on the **tens rod.** There is nothing to do on the ones rod because the ones place number is zero.

EXAMPLE: 2

1	After	+ 54	- 02	- 50	+ 11	= 13
54	ABACUS LOOKS LIKE					
- 02						
- 50						
11						
13						

Clear	Problem	Action
Step 1	+ 54	Move the **heaven bead down** to touch the beam on the **tens rod**. Move **4 earth beads up** to touch the beam on the **ones rod**.
Step 2	- 02	There is nothing to do on the tens rod because the tens place number is zero. Move **2 earth beads down** to touch the frame on the **ones rod**.
Step 3	- 50	Move the **heaven bead up** to touch the frame on the **tens rod**. There is nothing to do on the ones rod because the ones place number is zero.
Step 4	+ 11	Move **1 earth bead up** to touch the beam on the **tens rod**. Move **1 earth bead up** to touch the bead which is touching the beam on the **ones rod**.

EXAMPLE: 3

1	After	+ 05	+50	+41	- 55	= 41
05	ABACUS LOOKS LIKE					
50						
41						
- 55						
41						

SAI Speed Math Academy

LESSON 2 – SAMPLE PROBLEMS

1	2	3	4	5	6	7	8
			35	42	23	33	23
20	05	55	12	05	- 12	15	55
05	40	11	01	51	25	01	21
- 10	- 25	- 15	- 05	- 45	- 11	- 44	- 55

2:1

1	2	3	4	5	6	7	8
12	53	43	41	51	32	44	24
12	01	- 11	52	25	12	55	- 01
50	- 04	55	- 11	12	- 04	- 11	- 21
- 21	05	- 32	- 32	- 53	05	- 35	51

2:2

1	2	3	4	5	6	7	8
25	25	32	33	54	11	55	11
10	- 15	- 20	- 22	- 03	12	02	32
10	52	- 10	05	05	51	- 51	05
- 45	- 51	52	- 11	- 55	- 23	- 01	- 33

2:3

ATTENTION

- When there is no symbol in front of a number, then it means that the number is to be added.

- In the beginning, your child will try to **show** you the number instead of adding or subtracting the number.

LESSON 3 – INTRODUCING– NUMBERS 6 AND 7

NUMBERS USING EARTH BEADS AND HEAVEN BEADS

Set ONES PLACE NUMBERS on the ONES PLACE ROD of the abacus

6 = 5 + 1 Pack* one earth bead and one heaven bead together to touch the beam on the ones rod.

7 = 5 + 1 + 1 Pack* two earth beads and the heaven bead together to touch the beam on the ones rod.

Set TENS PLACE NUMBERS on the TENS PLACE ROD of the abacus

60 = 50 + 10 Pack* one earth bead and the heaven bead together to touch the beam on the tens rod.

70 = 50 + 10 + 10 Pack* two earth beads and the heaven bead together to touch the beam on the tens rod.

*Pack: Move both the heaven and earth beads together to touch the beam (adding) with proper fingers at the same time.

PRACTICE SETTING NUMBERS

After your child is comfortable setting and reading the above numbers proceed to introduce the numbers highlighted in the list below.

Activity 1: On the abacus, set any number from the above list and ask your child to read it.

Activity 2: Call out any number from the list and ask your child to set it on the abacus.

00	10	20	30	40	50	**60**	**70**
01	11	21	31	41	51	**61**	**71**
02	12	22	32	42	52	**62**	**72**
03	13	23	33	43	53	**63**	**73**
04	14	24	34	44	54	**64**	**74**
05	15	25	35	45	55	**65**	**75**
06	**16**	**26**	**36**	**46**	**56**	**66**	**76**
07	**17**	**27**	**37**	**47**	**57**	**67**	**77**

(Do these practices with any 10 numbers from the list.)

SAI Speed Math Academy

LESSON 3 – EXAMPLE

POINTS TO REMEMBER

Each heaven bead = 5

Each earth bead = 1

Fingering: Use appropriate fingers to move heaven and earth beads.

Zero: Always set abacus to zero by clearing all the beads away from the beam before starting each calculation.

Order of action: Set tens place number first and then set ones place number on the abacus.

EXAMPLE: 1

1	After	+ 26	+ 51	- 76	+ 66	= 67
26						
51	ABACUS					
- 76	LOOKS					
66	LIKE					
67						

Clear	Problem	Action
Step 1	+ 26	Move **2 earth beads up** to touch the beam on the **tens rod**. Move the **heaven bead down AND 1 earth bead up** to touch the beam on the **ones rod**.
Step 2	+ 51	Move the **heaven bead down** to touch the beam on the **tens rod**. Move **1 earth bead up** to touch the bead which is touching the beam on the **ones rod**.
Step 3	- 76	Move the **heaven bead up AND 2 earth beads down** to touch the frame on the **tens rod**. Move the **heaven bead up AND 1 earth bead down** to touch the frame on the **ones rod**.
Step 4	+ 66	Move the **heaven bead down AND 1 earth bead up** to touch the beam on the **tens rod**. Move the **heaven bead down AND 1 earth beads up** to touch the beam on the **ones rod**.

EXAMPLE: 2

1	After	+ 77	− 66	+ 56	− 67	= 00
77	ABACUS LOOKS LIKE					
− 66						
56						
− 67						
00						

Clear	Problem	Action
Step 1	+ 77	Move the **heaven bead down AND 2 earth beads up** to touch the beam on the **tens rod**. Move the **heaven bead down AND 2 earth beads up** to touch the beam on the **ones rod**.
Step 2	− 66	Move the **heaven bead up AND 1 earth bead down** to touch the frame on the **tens rod**. Move the **heaven bead up AND 1 earth bead down** to touch the frame on the **ones rod**.
Step 3	+ 56	Move the **heaven bead down** to touch the beam on the **tens rod**. Move the **heaven bead down AND 1 earth bead up** to touch the beam on the **ones rod**.
Step 4	− 67	Move the **heaven bead up AND 1 earth bead down** to touch the frame on the **tens rod**. Move the **heaven bead up AND 2 earth beads down** to touch the frame on the **ones rod**.

EXAMPLE: 3

1	After	+ 41	+55	+03	− 67	= 32
41	ABACUS LOOKS LIKE					
55						
03						
− 67						
32						

LESSON 3 – SAMPLE PROBLEMS

1	2	3	4	5	6	7	8
02	60	16	44	32	33	77	50
05	- 10	50	55	15	- 20	- 66	10
- 01	20	01	- 25	- 40	50	55	07

3:1

1	2	3	4	5	6	7	8
43	07	43	54	27	16	11	25
05	- 01	- 11	- 03	- 11	52	15	22
- 12	70	- 20	15	13	- 06	- 20	50
- 11	- 20	50	11	- 15	15	71	- 42

3:2

1	2	3	4	5	6	7	8
14	66	37	31	61	77	52	74
55	- 11	- 10	05	13	- 66	11	- 02
- 16	22	- 10	- 20	- 24	77	11	- 02
20	- 77	50	50	12	- 11	- 04	06

3:3

ATTENTION

- When there is no symbol in front of a number, then it means that the number is to be added.

- In the beginning children will try to **show** you the number instead of adding or subtracting a given number.

- With a combination number, children will forget to add or subtract by moving both heaven and earth beads together. They will work with one deck of beads and forget about the beads in the other deck.

www.abacus-math.com

LESSON 4 – INTRODUCING – NUMBERS 8 AND 9

Set UNITS PLACE NUMBERS on the UNITS PLACE ROD of the abacus

$8 = 5 + 1 + 1 + 1$ Pack three earth bead and one heaven bead together to touch the beam on the ones rod.

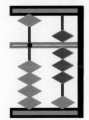

$9 = 5 + 1 + 1 + 1 + 1$ Pack four earth beads and the heaven bead together to touch the beam on the ones rod.

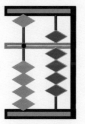

Set TENS PLACE NUMBERS on the TENS PLACE ROD of the abacus

$80 = 50+10+10+10$ Pack three earth bead and the heaven bead together to touch the beam on the tens rod.

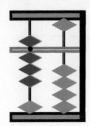

$90 = 50+10+10+10+10$ Pack four earth beads and the heaven bead together to touch the beam on the tens rod.

PRACTICE SETTING NUMBERS

After your child is comfortable setting and reading these numbers proceed to introduce the numbers highlighted in the list below.

Activity 1: On the abacus set any number from the above list and ask your child to read it.

Activity 2: Call out any number from the above list and ask your child to set it on the abacus.

(Do these practices with any 10 numbers from the above list.)

00	10	20	30	40	50	60	70	**80**	**90**
01	11	21	31	41	51	61	71	**81**	**91**
02	12	22	32	42	52	62	72	**82**	**92**
03	13	23	33	43	53	63	73	**83**	**93**
04	14	24	34	44	54	64	74	**84**	**94**
05	15	25	35	45	55	65	75	**85**	**95**
06	16	26	36	46	56	66	76	**86**	**96**
07	17	27	37	47	57	67	77	**87**	**97**
08	**18**	**28**	**38**	**48**	**58**	**68**	**78**	**88**	**98**
09	**19**	**29**	**39**	**49**	**59**	**69**	**79**	**89**	**99**

LESSON 4 – EXAMPLE

POINTS TO REMEMBER

Each heaven bead = 5

Each earth bead = 1

Fingering: Use appropriate fingers to move heaven and earth beads.

Zero: Always set abacus to zero by clearing all the beads away from the beam before starting each calculation.

Order of action: Set tens place number first and then set ones place number on the abacus.

EXAMPLE: 1

1	After	+ 28	+ 61	- 78	+ 87	= 98
28						
61	ABACUS					
- 78	LOOKS					
87	LIKE					
98						

Clear	Problem	Action
Step 1	+ 28	Move **2 earth beads up** to touch the beam on the **tens rod**. Move the **heaven bead down AND 3 earth beads up** to touch the beam on the **ones rod**.
Step 2	+ 61	Move the **heaven bead down AND 1 earth bead up** to touch the beam on the **tens rod**. Move **1 earth bead up** to touch the bead which is touching the beam on the **ones rod**.
Step 3	- 78	Move the **heaven bead up AND 2 earth beads down** to touch the frame on the **tens rod**. Move the **heaven bead up AND 3 earth beads down** to touch the frame on the **ones rod**.
Step 4	+ 87	Move the **heaven bead down AND 3 earth beads up** to touch the beam on the **tens rod**. Move the **heaven bead down AND 2 earth beads up** to touch the beam on the **ones rod**.

EXAMPLE: 2

1	After	+ 89	- 76	+ 85	- 98	= 00
89						
- 76	ABACUS					
85	LOOKS					
- 98	LIKE					
00						

Clear	Problem	Action
Step 1	+ 89	Move the **heaven bead down AND 3 earth beads up** to touch the beam on the **tens rod**. Move the **heaven bead down AND 4 earth beads up** to touch the beam on the **ones rod**.
Step 2	- 76	Move the **heaven bead up AND 2 earth beads down** to touch the frame on the **tens rod**. Move the **heaven bead up AND 1 earth bead down** to touch the frame on the **ones rod**.
Step 3	+ 85	Move the **heaven bead down AND 3 earth beads up** to touch the beam on the **tens rod**. Move the **heaven bead down** to touch the beam on the **ones rod**.
Step 4	- 98	Move the **heaven bead up AND 4 earth beads down** to touch the frame on the **tens rod**. Move the **heaven bead up AND 3 earth beads down** to touch the frame on the **ones rod**.

EXAMPLE: 3

1	After	+ 11	+88	-59	+50	= 90
11						
88	ABACUS					
- 59	LOOKS					
50	LIKE					
90						

SAI Speed Math Academy

LESSON 4 – SAMPLE PROBLEMS

1	*2*	*3*	*4*	*5*	*6*	*7*	*8*
02	03	11	29	32	58	22	99
05	01	67	50	55	40	12	- 88
01	05	11	- 66	10	- 70	55	77

4:1

1	*2*	*3*	*4*	*5*	*6*	*7*	*8*
04	07	01	07	08	50	20	50
- 03	- 01	01	01	01	30	10	30
05	- 05	06	- 02	- 02	- 10	- 20	- 10
01	08	01	03	01	- 10	70	- 20

4:2

1	*2*	*3*	*4*	*5*	*6*	*7*	*8*
85	96	32	31	49	59	22	91
- 20	- 11	67	08	- 06	20	51	07
33	03	- 98	- 22	55	20	05	- 63
01	11	52	80	- 98	- 08	- 58	54

4:3

ATTENTION

- When there is no symbol in front of a number, then it means that the number is to be added.

- In the beginning children will try to **show** you the number instead of adding or subtracting a given number.

- With a combination number, children will forget to add or subtract by moving both heaven and earth beads together. They will work with one deck of beads and forget about the beads in the other deck.

LESSON 5 – INTRODUCING – MIND MATH

YES! BELIEVE IT; YOUR CHILD IS READY TO START PRACTICING MIND MATH.

Because your child knows:

- There are four earth beads
- There is one heaven bead
- Tens place number has to be set on the tens place rod
- Ones place number has to be set on the ones place rod

What do you need to succeed?
Passion, patience, practice, tenacity and commitment.

MIND MATH – INSTRUCTIONS

MIND MATH is when a child tries to see the abacus in his/her mind and move the beads on the abacus in his/her mind to compute problems. Children have less mind clutter and are able to focus on their mental picture with clarity. Ask your child about his/her imaginary friends and s/he will be able to visualize and give you details about his/her imaginary friend. With little effort children will be able to pick up the skill and once they are comfortable they will prefer to work out problems in their mind rather than on the abacus.

STEP 1:

Sit comfortably and try to talk with children about school, friends, games or any of their interests.

Example: You can talk about their favorite friend.

Ask them what they did together the last time they met. What color clothing did they wear? What games did they play? What did they eat? What did they talk about, etc.?

While they are talking about their friend, ask them where they are seeing all these things. Ask them where they are visualizing their friend; after all, their friend is not in front of them. Ask where they are able to "think" about and "see" everything that they did together.

Children will definitely point to their head and say they are seeing their friend in their mind. If they are not able to answer you, then you can explain that even though their friend is not in front of them they can think of their friend inside their head and imagine that the friend is standing in front of them.

STEP 2:

Now ask them to see a big cloud, ball, or balloon in their favorite color in their head. (A cloud is the best choice.) This is to give them a back ground to hold their mental beam. If asked to just visualize a beam or abacus, children get confused and only see things in front of them rather than imagining the beam. Some children can imagine the beam without having to worry about any other object for support. Choose what fits their age and maturity level.

STEP 3:

At this point ask them where the cloud, ball, or balloon is hanging. Mostly they will show the space above their head. Now ask them to bring that colored cloud, ball, or balloon down on the table or in front of them. When children can imagine it on the table they are able to have better control of their mind abacus. But, some children are comfortable imagining their mind abacus in front of them.

STEP 4:

Now ask them to imagine drawing a small beam on the object they are visualizing in their favorite color. Tell them not to visualize the frame. On this beam ask them to imagine a tens rod and ones rod. It would help them if you build a mind abacus in front of you and show them where your beam, tens and ones rod are. This will help them to confidently build and see the beam in their mind, as everyone likes company while working.

STEP 5: Time to practice:

Ex: 1	Now you can ask them to:
01	Add one earth bead on the ones rod.
01	Add one more on the ones rod.
02	Then ask how many beads they have on the ones rod in their mind abacus.

Ex: 2	Now you can ask them to:
02	Add two earth beads on the ones rod.
- 01	Now minus one on the ones rod.
01	Then ask how many beads they have on the ones rod in their mind abacus.

Some children get it right away, some will need more effort. If child has difficulty imagining try any of the following steps.

Method 1: Keep the real abacus on the table in front of them. Clear the abacus and set it to zero. Now ask them to **imagine** moving the beads to compute.

Method 2: If the above method does not help, ask them to actually do the problem on the abacus and watch how the beads are moving. Repeat the problem a couple of times and then ask them to repeat Method 1. Once they are comfortable, try doing mind math problems without any visual help.

ATTENTION

VERY IMPORTANT:
- Child should visualize the beads only when they are in the game. Meaning, they should visualize only those beads that are touching the beam. Once beads are subtracted encourage them NOT to see those beads.

- As mind math is practiced visualizing rods is not necessary because children will *know* that left side of the beam is for tens place number and right side of the beam is for ones place number.

"WE DON'T NEED THEM, WE DON'T SEE THEM"

LESSON 5 – EXAMPLE

POINTS TO REMEMBER

- **VISUALIZE only the beads that are IN the game** If they are not in the game then you know that they are waiting to be included in the game, just like the substitute player of a team game like soccer or basketball.
- **FINGERS TO USE:** Use the same fingers to move your visual beads as on the physical abacus.

EXAMPLE: 1

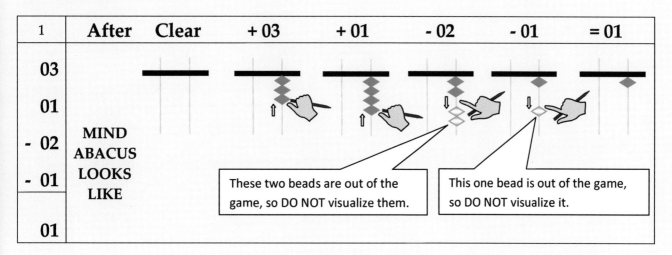

Clear	Problem	Action
Step 1	+ 03	There is nothing to do on tens rod. So, it stays empty. Move **3 earth beads up** to touch the beam on the **ones rod.**
Step 2	+ 01	There is nothing to do on tens rod. So, it stays empty. Move **1 earth bead up** to touch the bead which is touching the beam on the **ones rod.**
Step 3	- 02	There is nothing to do on tens rod. So, it stays empty. Move **2 earth beads away** from the ones rod. Do not see them anymore.
Step 4	- 01	There is nothing to do on tens rod. So, it stays empty. Move **1 earth bead away** from the ones rod. Do not see it anymore.

EXAMPLE: 2

1	After	Clear	+ 10	+ 50	- 10	+ 02	= 52
10							
50	MIND						
- 10	ABACUS LOOKS						
02	LIKE						
52							

This bead is out of the game, so DO NOT visualize it.

Clear	Problem	Action
Step 1	+ 10	Move **one earth bead up** to touch the beam on the **tens rod**. There is nothing to do on the ones rod, so it stays empty.
Step 2	+ 50	Move the **heaven bead down** to touch the beam on the **tens rod**. There is nothing to do on the ones rod, so it stays empty.
Step 3	- 10	Move **1 earth bead away** from the tens rod. Do not see it anymore. There is nothing to do on the ones rod, so it stays empty.
Step 4	+ 02	There is nothing to do on the tens rod. Move **2 earth beads up** to touch the beam on the **ones rod**.

EXAMPLE: 3

1	After	Clear	+ 25	+ 10	- 30	= 05
25						
10	MIND ABACUS					
- 30	LOOKS LIKE					
05						

These beads are out of the game, so DO NOT visualize them.

LESSON 5 – SAMPLE – MIND MATH

1	2	3	4	5	6	7	8
01	02	04	30	12	14	22	30
01	- 01	- 01	10	- 10	- 02	02	05
01	02	- 02	- 20	20	- 02	- 01	- 20

5:1

1	2	3	4	5	6	7	8
33	15	10	50	22	31	54	43
- 11	- 05	20	01	12	11	- 02	- 22
- 11	25	05	03	- 33	- 40	- 52	23

5:2

1	2	3	4	5	6	7	8
54	33	20	32	21	54	04	53
- 01	01	03	12	21	- 51	- 02	- 02
- 01	- 04	- 11	- 02	- 12	40	50	03
- 51	05	20	- 30	11	- 03	- 51	- 01

5:3

ATTENTION

As mind math is practiced visualizing rods is not necessary because children will *know* that left side of the beam is for tens place number and right side of the beam is for ones place number.

LESSON 6 – MIND MATH – Numbers 6 and 7

LESSON 6 – EXAMPLE

POINTS TO REMEMBER

- **VISUALIZE only the beads that are IN the game.** If they are not in the game then you know that it is waiting to be included in the game just (like the substitute player of a team game like soccer or basketball).
- **FINGERS TO USE:** Use the same fingers to move your visual beads as on the physical abacus.

EXAMPLE: 1

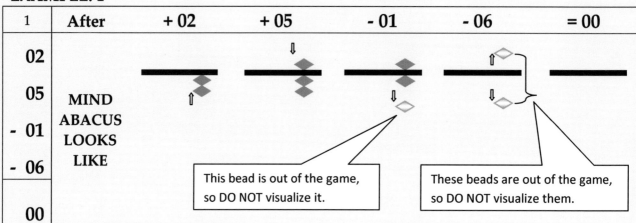

1	After	+ 02	+ 05	- 01	- 06	= 00
02						
05	MIND ABACUS LOOKS LIKE					
- 01						
- 06						
00						

Clear	Problem	Action
Step 1	+ 02	There is nothing to do on tens rod, so it stays empty. Move **two earth beads up** to touch the beam on the **ones rod**.
Step 2	+ 05	There is nothing to do on tens rod, so it stays empty. Move the **heaven bead down** to touch the beam on the **ones rod**.
Step 3	- 01	There is nothing to do on tens rod, so it stays empty. Move **one earth bead away** on the **ones rod**. Do not see it anymore.
Step 4	- 06	There is nothing to do on tens rod, so it stays empty. Move the **heaven bead and one earth bead away** from the beam on the **ones rod**. Do not see them anymore.

EXAMPLE: 2

1	After	+ 60	+ 14	- 02	- 70	= 02
60	MIND ABACUS LOOKS LIKE					
14						
- 02						
- 70						
02						

These two beads are out of the game, so DO NOT visualize them.

These beads are out of the game, so DO NOT visualize them.

Clear	Problem	Action
Step 1	+ 60	Move the **heaven bead down and one earth bead up** to touch the beam on the **tens rod**. There is nothing to do on ones rod. So, it stays empty
Step 2	+ 14	Move **one earth bead up** on the **tens rod**. Move **4 earth beads up** to touch the beam on the **ones rod**.
Step 3	- 02	There is nothing to do on tens rod. Move **2 earth beads away on the ones rod. Do not see them anymore.**
Step 4	- 70	Move **heaven bead and two earth beads away** from the beam. **Do not see them anymore.** There is nothing to do on ones rod.

EXAMPLE: 3

1	After	+ 75	- 60	+ 52	= 67
75	MIND ABACUS LOOKS LIKE				
- 60					
52					
67					

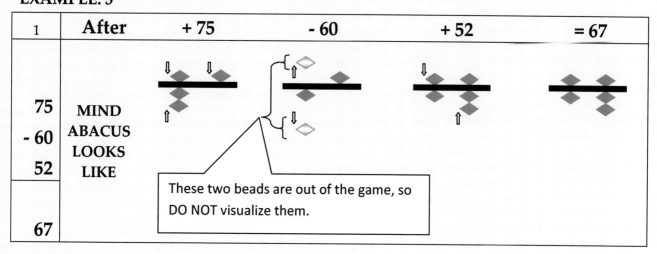

These two beads are out of the game, so DO NOT visualize them.

SAI Speed Math Academy

LESSON 6 – SAMPLE PROBLEMS

	1	2	3	4	5	6	7	8	
	05	06	05	10	70	10	10	06	
	01	01	02	60	- 20	60	02	20	6:1
	10	20	- 06	05	15	- 50	50	50	

	1	2	3	4	5	6	7	8	
	11	20	07	33	24	51	62	34	
	05	50	50	- 21	50	06	12	- 12	6:2
	11	03	10	50	- 13	- 07	- 04	25	

	1	2	3	4	5	6	7	8	
	04	03	06	42	40	10	45	31	
	- 02	01	11	05	- 30	15	01	- 20	6:3
	- 01	- 02	- 15	- 30	10	50	- 30	15	
	05	05	50	50	50	- 15	50	- 26	

ATTENTION

As mind math is practiced visualizing rods is not necessary because children will *know* that left side of the beam is for tens place number and right side of the beam is for ones place number.

LESSON 7 – MIND MATH – Numbers 8 and 9

LESSON 7 – EXAMPLE

POINTS TO REMEMBER

- **VISUALIZE only the beads that are IN the game.** If they are not in the game then you know that it is waiting to be included in the game (just like the substitute player of a team game like soccer or basketball).
- **FINGERS TO USE:** Use the same fingers to move your visual beads as on the physical abacus.

EXAMPLE: 1

1	After	+ 07	+ 02	- 08	+ 07	= 08
07	MIND ABACUS LOOKS LIKE					
02						
- 08						
07						
08						

These beads are out of the game, so DO NOT visualize them.

Clear	Problem	Action
Step 1	+ 07	There is nothing to do on tens rod, so it stays empty. Move the **heaven bead and two earth beads** to touch the beam on the **ones rod**.
Step 2	+ 02	There is nothing to do on tens rod, so it stays empty. Move the **2 earth beads up** on the **ones rod**.
Step 3	- 08	There is nothing to do on tens rod, so it stays empty. Move the **heaven bead and three earth beads away from the beam** on the ones rod. **Do not see them anymore.**
Step 4	+ 07	There is nothing to do on tens rod, so it stays empty. Move the **heaven bead and two earth beads** to touch the beam on the **ones rod**.

SAI Speed Math Academy

EXAMPLE: 2

1	After	+ 80	+ 14	+ 05	- 09	= 90
80	MIND ABACUS LOOKS LIKE					
14						
05						
- 09						
90						

These beads are out of the game, so DO NOT visualize them.

Clear	Problem	Action
Step 1	+ 80	Move **heaven bead down and three earth bead up** to touch the beam on the **tens rod**. There is nothing to do on ones rod. So, it stays empty
Step 2	+ 14	Move **one earth bead up** on the **tens rod**. Move **4 earth beads up** to touch the beam on the **ones rod**.
Step 3	+ 05	There is nothing to do on tens rod. Move the **heaven bead down** to touch the beam on the **ones rod**.
Step 4	- 09	There is nothing to do on tens rod. Move the **heaven bead and four earth beads away from the beam** on the **ones rod**. Do not see them anymore.

EXAMPLE: 3

1	After	+ 64	- 11	+ 45	= 98
64	MIND ABACUS LOOKS LIKE				
- 11					
45					
98					

These two beads are out of the game, so DO NOT visualize them.

Pg 40   www.abacus-math.com

LESSON 7 – SAMPLE PROBLEMS

1	*2*	*3*	*4*	*5*	*6*	*7*	*8*
06	05	04	03	40	30	52	09
01	02	50	15	- 10	50	30	- 06
01	02	05	10	50	10	02	05

7:1

1	*2*	*3*	*4*	*5*	*6*	*7*	*8*
15	24	60	45	69	13	27	44
02	05	30	02	- 14	11	- 11	55
01	50	- 70	02	40	05	02	- 08

7:2

1	*2*	*3*	*4*	*5*	*6*	*7*	*8*
09	14	16	51	22	05	66	54
10	05	02	20	15	12	- 11	- 03
20	- 11	- 03	- 11	51	12	33	40
- 30	30	04	20	11	- 09	- 80	07

7:3

ATTENTION

As mind math is practiced visualizing rods is not necessary because children will *know* that left side of the beam is for tens place number and right side of the beam is for ones place number.

LESSON 8 – SMALL FRIEND COMBINATIONS

SMALL FRIEND – COMBINATIONS OF 5

1 and 4 are friends of 5

1 + 4 = 5   4 + 1 = 5

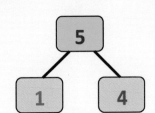

2 and 3 are friends of 5

2 + 3 = 5   3 + 2 = 5

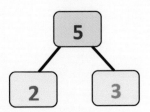

NUMBER SENTENCE FOR COMBINATIONS OF 5

Addition Facts	Subtraction Facts
1 + 4 = 5	5 − 1 = 4
4 + 1 = 5	5 − 4 = 1
2 + 3 = 5	5 − 2 = 3
3 + 2 = 5	5 − 3 = 2

1 and 4 are small friends, because together they make 5.
2 and 3 are small friends, because together they make 5.

POINTS TO REMEMBER

Make sure student understands the relation between small friend combination numbers. Soroban is based on these facts. All the concepts introduced in LEVEL – 1 will be using these small friend relationship facts to attain the desired results of adding or subtracting numbers 1 to 4 by manipulating the beads in a certain order.

The small friends concepts introduced in this level are like foundation for a building, thorough understanding of these facts is vital for their success with 10 exchange concepts, multiplication and division taught in higher levels. So, please make sure that students understand LEVEL – 1 concept very well.

EXTRA PRACTICE

- Quiz your child by asking "Who are friends of 5?" Answer: 1 & 4 are friends of 5. 2 & 3 are friends of 5.
- Ask "Why are 1 & 4 friends of 5 and why are 2 & 3 friends of 5?" Their answer should be something like; "Because 1 & 4 together make 5 and 2 & 3 together make 5".
- Confuse them by asking if 1 & 3 are friends of 5 or if 2 & 4 are friends of 5. The goal is for your child to know the relationship between the numbers. So, his/her answer has to be "no".

LESSON 9 – INTRODUCING +1 CONCEPT

LESSON 9 – EXAMPLE

CONCEPT OF THE WEEK

TO ADD = ADD 5, LESS SMALL FRIEND   **+ 1 = + 5 – 4**

EXAMPLE: 1

1	After	+ 04	+ 01 (+5 – 4)		= 05
04 01 05	ABACUS LOOKS LIKE	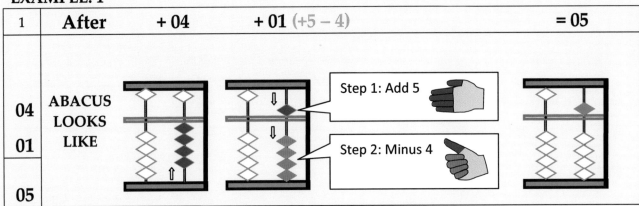			

Clear	Problem	Action
Step 1	+ 04	There is nothing to do on the tens rod because tens place number is zero. Move all **four earth beads up** to touch the beam on the **ones rod**.
Step 2	+ 01	There is nothing to do on the tens rod because tens place number is zero. Now we need to +1 on the ones rod, however we do not have enough earth beads to +1 on the ones rod. **When you want to +1 and you do not have enough beads:** Use +1 = +5 – 4 Small Friend formula to do your calculations. **Step 1: Add 5** – Move the heaven bead down to touch the beam on the ones rod. *(We know that there is a one in the five (1 + 4 = 5), so let us get help from 5 by including it in the game.* *But, we are supposed to +1, and when we +5 it means we have 4 more than what we need.* *So, now we have to send 4 away from our game.)* **Step 2: Minus 4** – Move all four earth beads away from the beam on the ones rod. *(When we do + 5 and – 4, we get to keep 1 in our game.)* +5 – 4 = +1

EXAMPLE: 2

| 1 | After | + 44 | (+50 − 40) + 15 | = 59 |

44	ABACUS LOOKS LIKE	Step 1: Add 50
15		Step 2: Minus 40
59		

Clear	Problem	Action
Step 1	+ 44	Move all **four earth beads up** to touch the beam on the **tens rod**. Move all **four earth beads up** to touch the beam on the **ones rod**.
Step 2	+ 15	Now we need to +10 on the tens rod, however we do not have enough earth beads to +10 on the tens rod. **When you want to +10 and you do not have enough beads:** Use +10 = +50 − 40 Small Friend formula to do your calculations. *(This can be taught as using the same bead movement as for +1, but on the tens rod.)* **Step 1: Add 50** Move the heaven bead down to touch the beam on the tens rod. **Step 2: Minus 40** Move all four earth beads down to touch the frame on the tens rod. Move the **heaven bead down** to touch the beam on the **ones rod**.

ATTENTION

- When doing +1, students will do +5 and forget to do − 4.

- When doing +5, students will also try to do − 4.

- Ask students to say the formula while they use it. This makes it easy for them to understand and follow through with all the steps.

SAI Speed Math Academy

POINTS TO REMEMBER

The row below consist of sample problems to introduce +1 = + 5 – 4 formula. Explain to your child when and how to use the formula. Use your fingers to give your child an idea about the heaven bead (use five fingers to represent five beads hiding inside the heaven bead). Work with the sample problems until your child understands the formula and that the formulas are to be used ONLY when there are not enough beads to add or subtract.

LESSON 9 – SAMPLE PROBLEMS

TO INTRODUCE +1 = + 5 – 4 FORMULA

Work these problems a few times to study and understand the concept and the relation between the beads moved.

1	2	3	4	5	6	7	8
	02	03		20	30		33
04	02	01	40	20	10	44	11
01	01	01	10	10	10	11	01

SAMPLE WORK

1	2	3	4	5	6	7	8
14	64	44	32	43	24	89	39
31	11	51	12	11	21	- 55	- 15
14	- 70	- 80	11	21	12	11	71

1	2	3	4	5	6	7	8
62	29	48	34	79	59	75	21
- 50	60	11	15	- 65	40	- 15	22
22	- 75	- 05	10	31	- 75	24	11
51	51	21	- 56	12	11	11	11

LESSON 10 – INTRODUCING –1 CONCEPT

LESSON 10 – EXAMPLE

CONCEPT OF THE WEEK

TO MINUS = MINUS 5, ADD SMALL FRIEND $-1 = -5 + 4$

EXAMPLE: 1

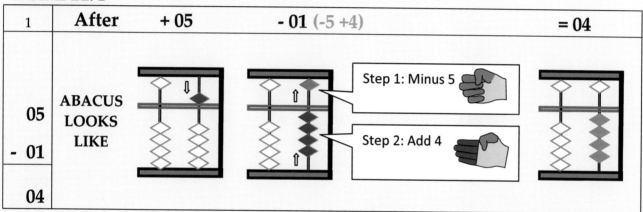

1	After	+ 05	- 01 (-5 +4)	= 04
05 - 01 04	ABACUS LOOKS LIKE		Step 1: Minus 5 Step 2: Add 4	

Clear	Problem	Action
Step 1	+ 05	There is nothing to do on the tens rod because tens place number is zero. Move the **heaven bead down** to touch the beam on the **ones rod.**
Step 2	- 01	There is nothing to do on the tens rod because tens place number is zero. Now we need to -1 on the ones rod, however we do not have enough earth beads to -1 on the ones rod. **When you want to -1 and you do not have enough beads:** Use $-1 = -5 +4$ Small Friend formula to do your calculations. **Step 1: Minus 5** – Move the heaven bead up to touch the frame on the ones rod. *(We know that there is a one in the five (1 + 4 = 5), so let us get help from 5 by sending it out of the game (– 5).* *We are supposed to – 1, but we did – 5, which means we have sent away 4 more than what we should have. So, now we have to add 4 into our game.)* **Step 2: Add 4** – Move all four earth beads up to touch the beam on the ones rod. *(When we do – 5 and + 4, we get to keep 1 out of game.)* $-5 + 4 = -1$

EXAMPLE: 2

		After	+ 56		(-50 +40) - 15	= 41
1						
56	ABACUS LOOKS LIKE			Step 1: Minus 50		
- 15				Step 2: Add 40		
41						

Clear	Problem	Action
Step 1	+ 56	Move the **heaven bead down** to touch the beam on the **tens rod**. Move the **heaven bead down** and **one earth beads up** to touch the beam on the **ones rod**.
Step 2	- 15	Now we need to – 10 on the tens rod, however we do not have enough earth beads to do the – 10 on the tens rod. **When you want to – 10 and you do not have enough beads:** Use – 10 = – 50 + 40 **Small Friend formula** to do your calculations. *(This can be taught as using the same bead movement as for – 1, but on the tens rod.)* **Step 1: Minus 50** - Move the heaven bead up to touch the frame on the tens rod. **Step 2: Add 40** – Move all four earth beads up to touch the beam on the tens rod. Move the **heaven bead up** to touch the frame on the **ones rod**.

ATTENTION

- When doing – 1, students will do –5 and forget to do +4.

- When doing –5, students will also try to do +4.

- Ask students to say the formula while they use it. This makes it easy for them to understand and follow through with all the steps.

www.abacus-math.com

POINTS TO REMEMBER

The rows below consist of sample problems to introduce –1 = – 5 + 4 formula. Explain to your child when and how to use the formula.   Use your fingers to give your child an idea about the heaven bead (five fingers as five beads hiding inside the heaven bead). Work with the sample problems until your child understands the formula and that the formulas are to be used ONLY when there are not enough beads to add or subtract.

LESSON 10 – SAMPLE PROBLEMS

TO INTRODUCE –1 = –5 + 4 FORMULA

Work these problems a few times to study and understand the concept and the relation between the beads moved.

1	2	3	4	5	6	7	8
05	56	55	04	60	44	34	75
- 01	- 11	- 11	01	- 10	11	11	- 21
			- 01	- 10	- 11	- 01	- 11

1	2	3	4	5	6	7	8
02	30	23	69	43	55	44	29
02	10	11	- 12	11	- 11	51	- 14
01	10	11	- 12	- 10	51	- 81	- 11
- 01	- 10	- 01	- 11	51	- 41	- 14	- 04

SAMPLE WORK

1	2	3	4	5	6	7	8
85	85	59	15	65	26	51	85
10	- 01	- 17	51	- 11	11	03	- 31
- 51	- 33	16	- 11	01	- 22	- 10	- 10
- 11	- 11	- 15	- 11	- 51	- 11	01	51

LESSON 11 – INTRODUCING +2 CONCEPT AND DICTATION

Because your child knows:

- There are four earth beads
- There is one heaven bead
- Tens place number has to be set on the tens place rod
- Ones place number has to be set on the ones place rod
- And is able to do mind math

LESSON 11 – EXAMPLE

CONCEPT OF THE WEEK

TO ADD = ADD 5, LESS SMALL FRIEND   +2 = +5 – 3

EXAMPLE: 1

1	After	+ 04	+ 02 (+5 – 3)	= 06
04 02 06	ABACUS LOOKS LIKE		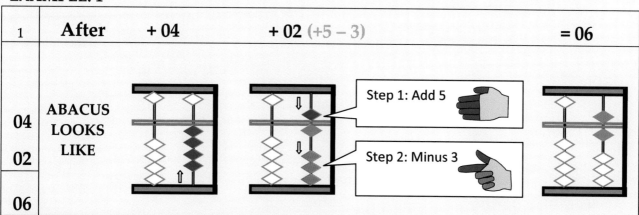	

Clear	Problem	Action
Step 1	+ 04	There is nothing to do on the tens rod because tens place number is zero. Move all **four earth beads up** to touch the beam on the **ones rod**.
Step 2	+ 02	There is nothing to do on the tens rod because tens place number is zero. Now we need to +2 on the ones rod, however we do not have enough earth beads to +2 on the ones rod. **When you want to +2 and you do not have enough beads:** Use +2 = +5 – 3 Small Friend formula to do your calculations. <u>**Step 1: Add 5**</u> – Move the heaven bead down to touch the beam on the ones rod. *(We know that there is a two in the five (2 + 3 = 5), so let us get help from 5 by including it in the game.* *But, we are supposed to +2, and when we +5 it means we have 3 more than what we need.* *So, now we have to send 3 away from our game.)* <u>**Step 2: Minus 3**</u> – Move three earth beads away from the beam on the ones rod. *(When we do + 5 and – 3, we get to keep 2 in our game.)* +5 – 3 = +2

EXAMPLE: 2

1	After	+ 33	(+50 − 30) + 25	= 58
33 25 58	ABACUS LOOKS LIKE	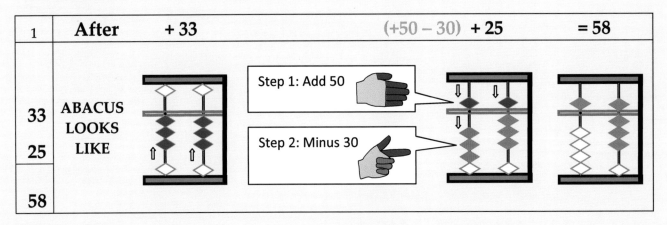		

Clear	Problem	Action
Step 1	+ 33	Move **three earth beads up** to touch the beam on the **tens rod**. Move **three earth beads up** to touch the beam on the **ones rod**.
Step 2	+ 25	Now we need to +20 on the tens rod, however we do not have enough earth beads to +20 on the tens rod. **When you want to +20 and you do not have enough beads:** Use +20 = +50 − 30 Small Friend formula to do your calculations. *(This can be taught as using the same bead movement as for +2, but on the tens rod.)* **Step 1: Add 50** Move the heaven bead down to touch the beam on the tens rod. **Step 2: Minus 30** Move three earth beads down to touch the frame on the tens rod. Move the **heaven bead down** to touch the beam on the **ones rod**.

ATTENTION

- When doing +2, students will do +5 and forget to do −3.

- When doing +5, students will also try to do −3.

- Ask students to **say** the formula while they use it. This makes it easy for them to understand and follow through with all the steps.

Pg 52   www.abacus-math.com

POINTS TO REMEMBER

First row is sample problems to introduce +2 = + 5 – 3 formula. Explain to your child when and how to use the formula. Use your hand and fingers to give your child the idea about the heaven bead (with five beads hiding inside). Work with the sample problems until your child understands the formula and that the formulas are to be used ONLY when there are not enough beads to add or subtract.

LESSON 11 – SAMPLE PROBLEMS

TO INTRODUCE +2 = + 5 – 3 FORMULA

Work these problems a few times to study and understand the concept and the relation between the beads moved.

1	2	3	4	5	6	7	8
		02	20			13	14
03	04	02	10	44	34	21	22
02	02	02	20	12	22	52	23

1	2	3	4	5	6	7	8
22	13	34	11	24	33	23	32
21	22	12	22	20	12	22	25
22	22	20	25	25	22	22	42

SAMPLE WORK

1	2	3	4	5	6	7	8
74	49	35	44	20	53	37	64
21	- 15	20	52	24	12	22	15
- 51	22	- 11	- 66	- 30	21	- 15	- 56
25	- 15	52	25	52	- 35	21	22

SAI Speed Math Academy

WEEK 11 – DICTATION

DICTATION: Dictation is when teacher or parent calls out a series of numbers and the child listens to the numbers and does the calculation in mind or on the abacus.

HOW TO GIVE A DICTATION? Choose any set of problems from the mind math part and call out the numbers to the child. Once the child is comfortable working on those problems, you can give more challenging ones from the 'use abacus' section of that week.

TWO WAYS OF WORKING ON DICTATION: As the child hears you call out the numbers, they can calculate it in two ways:

1. **On Abacus:** Child listens to the numbers and performs the calculation on the abacus. This is easy because s/he has a tool to work on and the only challenge is recognizing the place value of the numbers they are given and to add or subtract them on their corresponding rods.

2. **In Mind:** Child listens to the numbers and performs the calculation on the abacus in mind. This is more challenging because s/he has to listen to the numbers and then set them on his/her mind abacus. Then, s/he turns his/her attention back to you, while still holding the first number in mind, and listens to the second number along with the operation command (plus or minus), taking it back to their mind abacus, and working on it. They have to do all this while recognizing the place value of the numbers and adding or subtracting them on their corresponding rods.

   In the beginning, it is better to give two numbers and then ask them to give you the answer. Slowly, you can add more numbers. With more numbers the challenge is to concentrate for longer durations. Success with dictation will definitely increase their self confidence.

Some children are able to do dictation problems right away, some will need more effort. If child has difficulty imagining try any of the following steps.

Method 1: Keep the real abacus on the table in front of them. Clear the abacus and set it to zero. Now ask them to **imagine** moving the beads to compute as they hear numbers being called out.

Method 2: If the above method does not help, ask them to actually do the problem on the abacus and watch how the beads are moving. Repeat the problem a couple of times and then ask them to repeat Method 1. Once they are comfortable, try dictation problems without any visual help.

LESSON 12 – INTRODUCING –2 CONCEPT

LESSON 12 – EXAMPLE

CONCEPT OF THE WEEK

TO MINUS = MINUS 5, ADD SMALL FRIEND $-2 = -5 + 3$

EXAMPLE: 1

1	After	+ 05	- 02 (-5 +3)		= 03
05 - 02 03	ABACUS LOOKS LIKE		Step 1: Minus 5 Step 2: Add 3		

Clear	Problem	Action
Step 1	+ 05	There is nothing to do on the tens rod because tens place number is zero. Move the **heaven bead down** to touch the beam on the **ones rod**.
Step 2	- 02	There is nothing to do on the tens rod because tens place number is zero. Now we need to – 2 on the ones rod, however we do not have enough earth beads to – 2 on the ones rod. **When you want to – 2 and you do not have enough beads:** Use $-2 = -5 + 3$ Small Friend formula to do your calculations. **Step 1: Minus 5** – Move the heaven bead up to touch the frame on the ones rod. *(We know that there is a two in the five (2 + 3 = 5), so let us get help from 5 by sending it out of the game (– 5).* *We are supposed to – 2, but we did – 5, which means we have sent away 3 more than what we should have. So, now we have to add 3 into our game.)* **Step 2: Add 3** – Move three earth beads up to touch the beam on the ones rod. *(When we do – 5 and + 3, we get to keep 2 out of game.)* $-5 + 3 = -2$

EXAMPLE: 2

1		After	+ 67		(-50 +30) - 25	= 42
67 - 25 42	ABACUS LOOKS LIKE			Step 1: Minus 50 Step 2: Add 30		

Clear	Problem	Action
Step 1	+ 67	Move the **heaven bead down** and **one earth beads up** to touch the beam on the **tens rod**. Move the **heaven bead down** and **two earth beads up** to touch the beam on the **ones rod**.
Step 2	- 25	Now we need to – 20 on the tens rod, however we do not have enough earth beads to do the – 20 on the tens rod. **When you want to – 20 and you do not have enough beads:** Use – 20 = – 50 +30 Small Friend formula to do your calculations. *(This can be taught as using the same bead movement as for – 2, but on the tens rod.)* **Step 1: Minus 50** - Move the heaven bead up to touch the frame on the tens rod. **Step 2: Add 30** – Move three earth beads up on the tens rod. Move the **heaven bead up** to touch the frame on the **ones rod**.

ATTENTION

- When doing – 2, students will do –5 and forget to do +3.

- When doing –5, students will also try to do +3.

- Ask students to **say** the formula while they use it. This makes it easy for them to understand and follow through with all the steps.

POINTS TO REMEMBER

First row is sample problems to introduce –2 = – 5 + 3 formula. Explain to your child when and how to use the formula. Use your hand and fingers to give your child the idea about the heaven bead (with five beads hiding inside). Work with the sample problems until your child understands the formula and that the formulas are to be used ONLY when there are not enough beads to add or subtract.

LESSON 12 – SAMPLE PROBLEMS

TO INTRODUCE –2 = – 5 + 3 formula

Work these problems a few times to study and understand the concept and the relation between the beads moved.

1	2	3	4	5	6	7	8
05	06	55	56	56	67	85	74
- 02	- 02	- 22	- 22	- 12	- 22	- 20	11
					- 22	- 25	- 52

12:1

1	2	3	4	5	6	7	8
30	20	35	45	66	76	89	34
20	20	24	- 12	- 22	- 22	- 22	11
10	10	- 25	21	- 22	- 11	- 22	- 02
- 20	- 20	- 33	- 20	- 22	52	- 22	25

12:2

SAMPLE WORK

1	2	3	4	5	6	7	8
25	15	96	67	88	53	46	75
71	50	- 30	- 21	- 33	02	10	- 62
- 42	- 02	- 21	- 15	- 52	- 11	- 22	30
- 10	20	- 42	27	- 03	25	- 32	12

12:3

SAI Speed Math Academy

LESSON 13 – INTRODUCING +3 CONCEPT

LESSON 13 – EXAMPLE

CONCEPT OF THE WEEK

<u>TO ADD</u> = ADD 5, LESS SMALL FRIEND   <u>+ 3 = + 5 – 2</u>

EXAMPLE: 1

1		After	+ 03	+ 03 (+5 – 2)		= 06
03	ABACUS LOOKS LIKE					
03						
06						

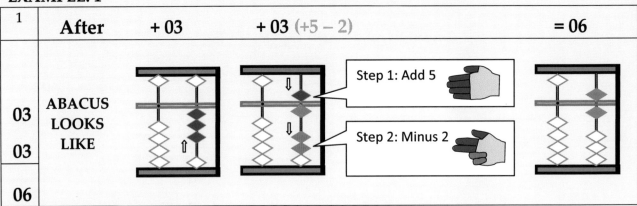

Clear	Problem	Action
Step 1	+ 03	There is nothing to do on the tens rod because tens place number is zero. Move **three earth beads up** to touch the beam on the **ones rod**.
Step 2	+ 03	There is nothing to do on the tens rod because tens place number is zero. Now we need to +3 on the ones rod, however we do not have enough earth beads to +3 on the ones rod. **When you want to +3 and you do not have enough beads:** Use +3 = +5 – 2 **Small Friend formula to do your calculations.** **<u>Step 1: Add 5</u>** – Move the heaven bead down to touch the beam on the ones rod. *(We know that there is a three in the five (3 + 2 = 5), so let us get help from 5 by including it in the game.* *But, we are supposed to +3, and when we +5 it means we have 2 more than what we need. So, now we have to send 2 away from our game.)* **<u>Step 2: Minus 2</u>** – Move two earth beads away from the beam on the ones rod. *(When we do + 5 and – 2, we get to keep 3 in our game.)* +5 – 2 = +3

EXAMPLE: 2

1		**After**	+ 42	(+50 − 20) + 33	= 75
42	ABACUS LOOKS LIKE			Step 1: Add 50	
33				Step 2: Minus 20	
75					

Clear	Problem	Action
Step 1	+ 42	Move all the **four earth beads up** to touch the beam on the **tens rod**. Move **two earth beads up** to touch the beam on the **ones rod**.
Step 2	+ 33	Now we need to +30 on the tens rod, however we do not have enough earth beads to +30 on the tens rod. **When you want to +30 and you do not have enough beads:** Use +30 = +50 − 20 **Small Friend formula to do your calculations.** *(This can be taught as using the same bead movement as for +3, but on the tens rod.)* **Step 1: Add 50** Move the heaven bead down to touch the beam on the tens rod. **Step 2: Minus 20** Move two earth beads down to touch the frame on the tens rod. Need to +3 on the ones rod and we do not have enough beads so, use the formula. **Step 1: Add 5** – Move the heaven bead down to touch the beam on the ones rod. **Step 2: Minus 2** – Move two earth beads away from the beam on the ones rod.

ATTENTION
- When doing +3, students will do +5 and forget to do − 2.
- When doing +5, students will also try to do − 2.
- Students most likely will confuse between +3 and +2 formula, so care has to be taken to make them understand that 2 and 3 are friends of 5 and **they help each other**.

POINTS TO REMEMBER

First row is sample problems to introduce +3 = + 5 – 2 formula. Explain to your child when and how to use the formula. Use your hand and fingers to give your child the idea about the heaven bead (with five beads hiding inside). Work with the sample problems until your child understands the formula and that the formulas are to be used ONLY when there are not enough beads to add or subtract.

LESSON 13 – SAMPLE PROBLEMS

TO INTRODUCE +3 = + 5 – 2 FORMULA

Work these problems a few times to study and understand the concept and the relation between the beads moved.

1	2	3	4	5	6	7	8
02	03	04	21	32	46	24	31
03	03	03	33	35	33	33	33

1	2	3	4	5	6	7	8
11	42	34	23	44	11	25	14
33	03	33	31	30	13	32	33
33	- 11	- 55	33	- 51	30	- 15	30
22	53	63	- 77	33	03	33	12

SAMPLE WORK

1	2	3	4	5	6	7	8
24	54	13	51	28	75	55	25
53	13	12	25	31	- 22	- 11	- 21
- 15	- 25	33	- 52	- 14	- 23	33	32
03	31	20	35	30	34	- 25	33

LESSON 14 – INTRODUCING –3 CONCEPT

LESSON 14 – EXAMPLE

CONCEPT OF THE WEEK

TO MINUS = MINUS 5, ADD SMALL FRIEND $-3 = -5 + 2$

EXAMPLE: 1

1		After	+ 07	- 03 (-5 +2)		= 04
07 - 03 04	ABACUS LOOKS LIKE					

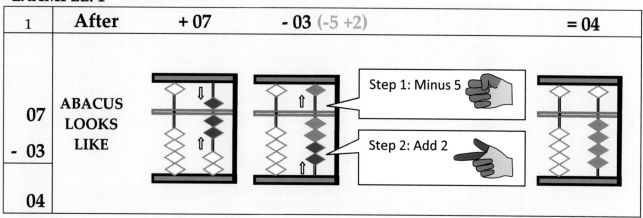

Clear	Problem	Action
Step 1	+ 07	There is nothing to do on the tens rod because tens place number is zero. Move the **heaven bead down** and **two earth beads up** to touch the beam on the **ones rod.**
Step 2	- 03	There is nothing to do on the tens rod because tens place number is zero. Now we need to – 3 on the ones rod, however we do not have enough earth beads to – 3 on the ones rod. **When you want to – 3 and you do not have enough beads:** Use – 3 = – 5 +2 Small Friend formula to do your calculations. **Step 1: Minus 5** – Move the heaven bead up to touch the frame on the ones rod. (We know that there is a three in the five (3 + 2 = 5), so let us get help from 5 by sending it out of the game (– 5). We are supposed to – 3, but we did – 5, which means we have sent away 2 more than what we should have. So, now we have to add 2 into our game.) **Step 2: Add 2** – Move two earth beads up to touch the beam on the ones rod. (When we do – 5 and + 2, we get to keep 3out of game.) $-5 + 2 = -3$

EXAMPLE: 2

| | After | + 65 | (-50 +20) - 30 | = 35 |

	ABACUS LOOKS LIKE	
65		Step 1: Minus 50
- 30		Step 2: Add 20
35		

Clear	Problem	Action
Step 1	+ 65	Move the **heaven bead down** and **one earth bead up** to touch the beam on the **tens rod**. Move the **heaven bead down** to touch the beam on the **ones rod**.
Step 2	− 30	Now we need to − 30 on the tens rod, however we do not have enough earth beads to do the − 30 on the tens rod. **When you want to − 30 and you do not have enough beads:** Use − 30 = − 50 +20 Small Friend formula to do your calculations. (This can be taught as using the same bead movement as for − 3, but on the tens rod.) **Step 1: Minus 50** - Move the heaven bead up to touch the frame on the tens rod. **Step 2: Add 20** – Move two earth beads up on the tens rod. There is nothing to do on the ones rod.

ATTENTION

- When doing − 3, students will do − 5 and forget to do + 2.
- When doing − 5, students will also try to do + 2.
- Students most likely will confuse between − 3 and − 2 formula, so care has to be taken to make them understand that 2 and 3 are friends of 5 and **they help each other**.
- Ask students to **say** the formula while they use it. This makes it easy for them to understand and follow through with all the steps.

POINTS TO REMEMBER

First row is sample problems to introduce –3 = – 5 + 2 formula. Explain to your child when and how to use the formula.   Use your hand and fingers to give your child the idea about the heaven bead (with five beads hiding inside). Work with the sample problems until your child understands the formula and that the formulas are to be used ONLY when there are not enough beads to add or subtract.

LESSON 14 – SAMPLE PROBLEMS

TO INTRODUCE –3 = – 5 + 2 FORMULA

Work these problems a few times to study and understand the concept and the relation between the beads moved.

1	2	3	4	5	6	7	8
05	06	57	68	56	57	78	97
- 03	- 03	- 33	- 33	- 33	- 33	- 33	- 33
			- 03	- 03	- 03	- 33	- 33

1	2	3	4	5	6	7	8
02	03	30	20	67	49	22	86
02	03	30	35	- 30	- 02	23	- 33
01	- 01	10	10	- 03	10	11	- 30
- 03	- 03	- 30	- 35	35	- 35	- 53	55

SAMPLE WORK

1	2	3	4	5	6	7	8
65	75	52	81	42	79	54	15
- 31	- 53	13	15	12	- 13	12	61
22	35	- 30	- 22	- 33	- 53	- 11	- 53
- 13	- 53	- 15	- 33	25	82	- 35	32

LESSON 15 – INTRODUCING +4 CONCEPT

LESSON 15 – EXAMPLE

CONCEPT OF THE WEEK
TO ADD = ADD 5, LESS SMALL FRIEND   + 4 = + 5 – 1

EXAMPLE: 1

1	After	+ 03	+ 04 (+5 – 1)	= 07
03 04 07	ABACUS LOOKS LIKE	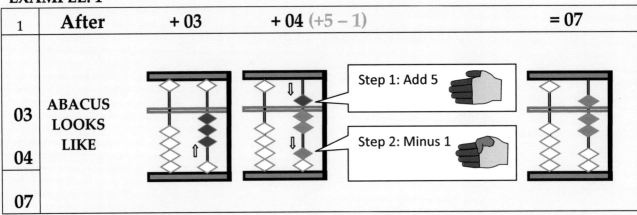		

Clear	Problem	Action
Step 1	+ 03	There is nothing to do on the tens rod because tens place number is zero. Move **three earth beads up** to touch the beam on the **ones rod**.
Step 2	+ 04	There is nothing to do on the tens rod because tens place number is zero. Now we need to +4 on the ones rod, however we do not have enough earth beads to +4 on the ones rod. **When you want to +4 and you do not have enough beads:** Use +4 = +5 – 1 Small Friend formula to do your calculations. **Step 1: Add 5** – Move the heaven bead down to touch the beam on the ones rod. (We know that there is a four in the five (4 + 1 = 5), so let us get help from 5 by including it in the game. But, we are supposed to +4, and when we +5 it means we have 1 more than what we need. So, now we have to send 1 away from our game.) **Step 2: Minus 1** – Move one earth bead away from the beam on the ones rod. (When we do + 5 and – 1, we get to keep 4 in our game.) +5 – 1 = +4

EXAMPLE: 2

1	After	+ 42	(+50 – 10) + 44	= 86
42	ABACUS LOOKS LIKE		Step 1: Add 50	
44			Step 2: Minus 10	
86				

Clear	Problem	Action
Step 1	+ 42	Move all the **four earth beads up** to touch the beam on the **tens rod**. Move **two earth beads up** to touch the beam on the **ones rod**.
Step 2	+ 44	Now we need to +40 on the tens rod, however we do not have enough earth beads to +40 on the tens rod. **When you want to +40 and you do not have enough beads:** Use +40 = +50 – 10 Small Friend formula to do your calculations. *(This can be taught as using the same bead movement as for +4, but on the tens rod.)* **Step 1: Add 50** Move the heaven bead down to touch the beam on the tens rod. **Step 2: Minus 10** Move one earth bead down to touch the frame on the tens rod. Need to +4 on the ones rod and we do not have enough beads so, use the formula. **Step 1: Add 5** – Move the heaven bead down to touch the beam on the ones rod. **Step 2: Minus 1** – Move one earth beads away from the beam on the ones rod.

ATTENTION

- When doing +4, students will do +5 and forget to do – 1.
- When doing +5, students will also try to do – 1.
- Students most likely will confuse between +1 and +4 formula, so care has to taken to make them understand that 4 and 1 are friends of 5 and **they help each other**.

POINTS TO REMEMBER

First row is sample problems to introduce +4 = + 5 – 1 formula. Explain to your child when and how to use the formula. Use your hand and fingers to give your child the idea about the heaven bead (with five beads hiding inside). Work with the sample problems until your child understands the formula and that the formulas are to be used ONLY when there are not enough beads to add or subtract.

LESSON 15 – SAMPLE PROBLEMS

TO INTRODUCE +4 = + 5 –1 FORMULA

Work these problems a few times to study and understand the concept and the relation between the beads moved.

1	2	3	4	5	6	7	8
02	03	04	12	24	33	44	14
04	04	04	44	44	44	45	44

1	2	3	4	5	6	7	8
23	11	12	20	51	01	34	25
44	44	12	20	44	54	40	44
- 20	- 13	54	45	- 62	04	04	- 55
41	45	- 25	- 61	14	- 59	- 53	44

SAMPLE WORK

1	2	3	4	5	6	7	8
11	34	22	67	82	21	75	31
45	- 21	42	- 23	11	54	14	34
11	45	14	54	- 62	- 32	- 55	- 22
- 62	- 51	- 78	- 47	44	44	14	45

LESSON 16 – INTRODUCING –4 CONCEPT

LESSON 16 – EXAMPLE

CONCEPT OF THE WEEK

TO MINUS = MINUS 5, ADD SMALL FRIEND $-4 = -5 + 1$

EXAMPLE: 1

1	After	+ 06	- 04 (-5 +1)	= 02
06 - 04 02	ABACUS LOOKS LIKE	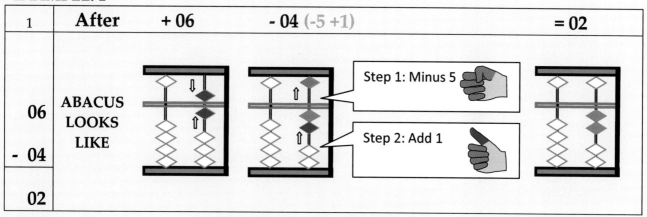		

Step 1: Minus 5

Step 2: Add 1

Clear	Problem	Action
Step 1	+ 06	There is nothing to do on the tens rod because tens place number is zero. Move the **heaven bead down** and **one earth bead up** to touch the beam on the **ones rod**.
Step 2	- 04	There is nothing to do on the tens rod because tens place number is zero. Now we need to – 4 on the ones rod, however we do not have enough earth beads to – 4 on the ones rod. **When you want to – 4 and you do not have enough beads:** Use – 4 = – 5 +1 **Small Friend** formula to do your calculations. **Step 1: Minus 5** – Move the heaven bead up to touch the frame on the ones rod. *(We know that there is a four in the five (4 + 1 = 5), so let us get help from 5 by sending it out of the game (– 5).* *We are supposed to – 4, but we did – 5, which means we have sent away 1 more than what we should have. So, now we have to add 1 into our game.)* **Step 2: Add 1** – Move one earth bead up to touch the beam on the ones rod. *(When we do – 5 and + 1, we get to keep 4 out of game.)* $-5 + 1 = -4$

EXAMPLE: 2

		After	+ 89	(−50 +10) − 44	= 45
	1				
	89	ABACUS LOOKS LIKE		Step 1: Minus 50	
	− 44			Step 2: Add 10	
	45				

Clear	Problem	Action
Step 1	+ 89	Move the **heaven bead down** and **three earth beads up** to touch the beam on the **tens rod**. Move the **heaven bead down** and **four earth beads up** to touch the beam on the **ones rod**.
Step 2	− 44	Now we need to − 40 on the tens rod, however we do not have enough earth beads to do the − 40 on the tens rod. **When you want to − 40 and you do not have enough beads:** Use − 40 = − 50 +10 Small Friend formula to do your calculations. (This can be taught as using the same bead movement as for − 4, but on the tens rod.) **Step 1: Minus 50** - Move the heaven bead up to touch the frame on the tens rod. **Step 2: Add 10** – Move one earth beads up on the tens rod. Move **four earth beads down** to touch the frame on the **ones rod**.

ATTENTION

- When doing − 4, students will do − 5 and forget to do + 1.
- When doing − 5, students will also try to do + 1.
- Students most likely will confuse between − 4 and − 1 formula, so care has to taken to make them understand that 4 and 1 are friends of 5 and **they help each other**.

POINTS TO REMEMBER

First row is sample problems to introduce –4 = – 5 + 1 formula. Explain to your child when and how to use the formula. Use your hand and fingers to give your child the idea about the heaven bead (with five beads hiding inside). Work with the sample problems until your child understands the formula and that the formulas are to be used ONLY when there are not enough beads to add or subtract.

LESSON 16 – SAMPLE PROBLEMS

TO INTRODUCE – 4 = – 5 + 1 FORMULA

Work these problems a few times to study and understand the concept and the relation between the beads moved.

1	2	3	4	5	6	7	8
						56	98
05	06	57	68	75	85	- 04	- 44
- 04	- 04	- 44	- 44	- 43	- 42	- 40	- 44

1	2	3	4	5	6	7	8
20	52	33	48	35	19	67	95
30	02	24	20	41	40	- 44	- 04
10	03	- 42	- 40	11	20	32	- 40
- 40	- 44	- 14	- 04	- 45	- 45	- 54	- 40

SAMPLE WORK

1	2	3	4	5	6	7	8
79	65	54	34	53	43	75	45
- 34	- 41	22	21	- 41	33	- 41	13
- 34	32	- 44	- 41	64	- 44	- 14	- 54
45	- 14	13	65	- 45	15	- 20	41

SAI Speed Math Academy

LESSON 17 – INTRODUCING – HUNDRED'S PLACE NUMBER

PLACE VALUE OF RODS

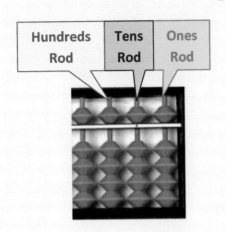

HUNDREDS ROD

The hundred's place number has to be set on the hundreds place rod.

USING FORMULAS ON HUNDRED'S ROD

When you have to add 1, 2, 3, 4 or subtract 1, 2, 3, 4 on the hundreds rod and you don't have enough beads, use the same small friends formula as used on the tens and ones place rod.

TO ADD	TO MINUS
+ 100 = + 500 – 400	– 100 = – 500 + 400
+ 200 = + 500 – 300	– 200 = – 500 + 300
+ 300 = + 500 – 200	– 300 = – 500 + 200
+ 400 = + 500 – 100	– 400 = – 500 + 100

LESSON 17 – EXAMPLE

EXAMPLE: 1

1	After	+ 400		+ 100 (+500 – 400) = 500
400 100 500	ABACUS LOOKS LIKE		Step 1: Add 500 Step 2: Minus 400	

Pg 70   www.abacus-math.com

EXAMPLE: 2

1	After	+ 500			− 100 (−500 + 400)	= 400
500 − 100 400	ABACUS LOOKS LIKE		Step 1: Minus 500 Step 2: Add 400			

EXAMPLE: 3

1	After	+ 565	− 342	+ 423	= 646
565 − 342 423 646	ABACUS LOOKS LIKE				

EXAMPLE: 4

1	After	+ 579	− 137	+ 243	= 685
579 − 137 243 685	ABACUS LOOKS LIKE				

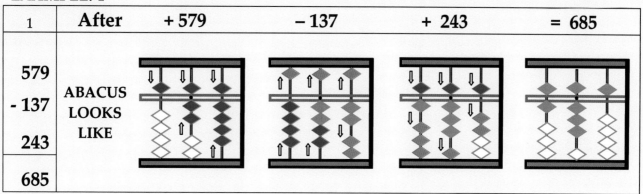

ATTENTION

- Be careful in reading a number and adding or subtracting on the correct place rod.
- Children get confused when a number has a zero in tens or ones place.

LESSON 17 – SAMPLE PROBLEMS

1	2	3	4	5	6	7	8
226	198	395	855	197	285	414	807
273	- 61	- 171	101	- 64	- 43	185	- 304
- 44	442	445	- 843	142	- 21	- 45	53
- 344	- 24	- 229	602	- 71	235	- 351	- 130
425	- 121	316	44	184	- 416	130	- 203

17:1

1	2	3	4	5	6	7	8
369	192	897	446	117	344	355	192
610	306	- 361	120	452	533	514	104
- 107	401	121	23	- 48	- 204	- 469	- 56
- 130	- 515	- 207	- 146	35	303	343	549
- 640	- 41	35	- 403	- 526	- 175	12	- 351

17:2

1	2	3	4	5	6	7	8
875	566	413	541	587	653	958	271
- 241	- 234	353	212	- 104	- 242	- 644	505
- 431	621	- 224	- 341	- 141	371	312	- 634
221	- 742	- 112	512	- 131	- 440	- 414	421
121	274	237	- 602	458	151	- 112	433
423	- 352	- 416	531	- 521	105	524	- 695

17:3

Pg 72

www.abacus-math.com

ANSWER KEY

LESSON 1

1	2	3	4	5	6	7	8	
04	20	34	11	30	04	02	01	1:1

1	2	3	4	5	6	7	8	
41	42	13	34	00	11	34	44	1:2

1	2	3	4	5	6	7	8	
22	22	32	12	23	10	01	03	1:3

LESSON 2

1	2	3	4	5	6	7	8	
15	20	51	43	53	25	05	44	2:1

1	2	3	4	5	6	7	8	
53	55	55	50	35	45	53	53	2:2

1	2	3	4	5	6	7	8	
00	11	54	05	01	51	05	15	2:3

LESSON 3

1	2	3	4	5	6	7	8	
06	70	67	74	07	63	66	67	3:1

1	2	3	4	5	6	7	8	
25	56	62	77	14	77	77	55	3:2

1	2	3	4	5	6	7	8	
73	00	67	66	62	77	70	76	3:3

LESSON 4

1	2	3	4	5	6	7	8	
08	09	89	13	97	28	89	88	4:1

1	2	3	4	5	6	7	8	
07	09	09	09	08	60	80	50	4:2

1	2	3	4	5	6	7	8	
99	99	53	97	00	91	20	89	4:3

LESSON 5

1	2	3	4	5	6	7	8	
03	03	01	20	22	10	23	15	5:1

1	2	3	4	5	6	7	8	
11	35	35	54	01	02	00	44	5:2

1	2	3	4	5	6	7	8	
01	35	32	12	41	40	01	53	5:3

LESSON 6

1	2	3	4	5	6	7	8	
16	27	01	75	65	20	62	76	6:1

1	2	3	4	5	6	7	8	
27	73	67	62	61	50	70	47	6:2

1	2	3	4	5	6	7	8	
06	07	52	67	70	60	66	00	6:3

LESSON 7

1	2	3	4	5	6	7	8	
08	09	59	28	80	90	84	08	7:1

1	2	3	4	5	6	7	8	
18	79	20	49	95	29	18	91	7:2

1	2	3	4	5	6	7	8	
09	38	19	80	99	20	08	98	7:3

LESSON 9

1	2	3	4	5	6	7	8	
05	05	05	50	50	50	55	45	9:1

SAMPLE WORK

1	2	3	4	5	6	7	8	
59	05	15	55	75	57	45	95	9:2

1	2	3	4	5	6	7	8	
85	65	75	03	57	35	95	65	9:3

LESSON 10

1	2	3	4	5	6	7	8	
04	45	44	04	40	44	44	43	10:1

1	2	3	4	5	6	7	8	
04	40	44	34	95	54	00	00	10:2

SAMPLE WORK

1	2	3	4	5	6	7	8	
33	40	43	44	04	04	45	95	10:3

LESSON 11

1	2	3	4	5	6	7	8	
05	06	06	50	56	56	86	59	11:1

1	2	3	4	5	6	7	8	
65	57	66	58	69	67	67	99	11:2

SAMPLE WORK

1	2	3	4	5	6	7	8	
69	41	96	55	66	51	65	45	11:3

LESSON 12

1	2	3	4	5	6	7	8	
03	04	33	34	44	23	40	33	12:1

1	2	3	4	5	6	7	8	
40	30	01	34	00	95	23	68	12:2

SAMPLE WORK

1	2	3	4	5	6	7	8	
44	83	03	58	00	69	02	55	12:3

LESSON 13

1	2	3	4	5	6	7	8	
05	06	07	54	67	79	57	64	13:1

1	2	3	4	5	6	7	8	
99	87	75	10	56	57	75	89	13:2

SAMPLE WORK

1	2	3	4	5	6	7	8	
65	73	78	59	75	64	52	69	13:3

LESSON 14

1	2	3	4	5	6	7	8	
02	03	24	32	20	21	12	31	14:1

1	2	3	4	5	6	7	8	
02	02	40	30	69	22	03	78	14:2

SAMPLE WORK

1	2	3	4	5	6	7	8	
43	04	20	41	46	95	20	55	14:3

LESSON 15

1	2	3	4	5	6	7	8	
06	07	08	56	68	77	89	58	15:1

1	2	3	4	5	6	7	8	
88	87	53	24	47	00	25	58	15:2

SAMPLE WORK

1	2	3	4	5	6	7	8	
05	07	00	51	75	87	48	88	15:3

LESSON 16

1	2	3	4	5	6	7	8	
01	02	13	24	32	43	12	10	16:1

1	2	3	4	5	6	7	8	
20	13	01	24	42	34	01	11	16:2

SAMPLE WORK

1	2	3	4	5	6	7	8	
56	42	45	79	31	47	00	45	16:3

LESSON 17

1	2	3	4	5	6	7	8	
536	434	756	759	388	40	333	223	17:1

1	2	3	4	5	6	7	8	
102	343	485	40	30	801	755	438	17:2

1	2	3	4	5	6	7	8	
968	133	251	853	148	598	624	301	17:3

ABOUT SAI SPEED MATH ACADEMY

One subject that is very important for success in this world, along with being able to read and write, is the knowledge of numbers. Math is one subject which requires proficiency from anyone who wants to achieve something in life. A strong foundation and a basic understanding of math is a must to mastering higher levels of math.

We, the family, best friends, and parents of children in elementary school, early on discovered that what our children were learning at school was not enough for them to master the basics of math. Teachers at school, with the resources they had, did the best they could. But, as parents, we had to do more to help them understand the relationship between numbers and basic functions of adding, subtracting, multiplying and dividing. Also, what made us cringe is the fact that our children's attitude towards more complex math was to say, "Oh, we are allowed to use a calculator in class". This did not sit well with us. Even though we did not have a specific system that we followed, each of us could do basic calculations in our minds without looking for a calculator. So, this made us want to do more for our children.

We started to look into the various methods that were available in the marketplace to help our children understand basic math and reduce their dependency on calculators. We came across soroban, a wonderful calculating tool from Japan. Soroban perfectly fits with the base-10 number system used at present and provides a systematic method to follow while calculating in one's mind.

This convinced us and within a short time we were able to work with fluency on the tool. The next step was to introduce it to our children, which we thought was going to be an easy task. It, however, was not. It was next to impossible to find the resources or the curriculum to help us introduce the tool in the correct order. Teaching all the concepts in one sitting and expecting children to apply them to the set of problems we gave them only made them push away the tool in frustration.

However, help comes to those who ask, and to those who are willing to work to achieve their goals. We came across a soroban teacher who helped us by giving us ideas and an outline of how soroban should be introduced. But, we still needed an actual worksheet to give our children to practice on. That is when we decided to come up with practice worksheets of our own design for our kids.

Slowly and steadily, practicing with the worksheets that we developed, our children started to get the idea and loved what they could do with a soroban. Soon we realized that they were better with mind math than we were.

Today, 6 years later, all our kids have completed their soroban training and are reaping the benefits of the hard work that they did over the years.

Now, although very happy, we were humbled at the number of requests we got from parents who wanted to know more about our curriculum. We had no way to share our new knowledge with them.

Now, through the introduction of our instruction book and workbooks, that has changed. We want to share everything we know with all the dedicated parents who are interested in teaching soroban to their children. This is our humble attempt to bring a systematic instruction manual and corresponding workbook to help introduce your children to soroban.

What started as a project to help our kids has grown over the years and we are fortunate to say that a number of children have benefitted learning with the same curriculum that we developed for our children.

Thank you for choosing our system to enhance your children's mathematical skills.

We love working on soroban and hope you do too!

List of SAI Speed Math Academy Publications

LEVEL – 1

Abacus Mind Math Instruction Book Level – 1: Step by Step Guide to Excel at Mind Math with Soroban, a Japanese Abacus
ISBN-13: 978-1941589007

Abacus Mind Math Level – 1 Workbook 1 of 2: Excel at Mind Math with Soroban, a Japanese Abacus
ISBN-13: 978-1941589014

Abacus Mind Math Level – 1 Workbook 2 of 2: Excel at Mind Math with Soroban, a Japanese Abacus
ISBN-13: 978-1941589021

LEVEL – 2

Abacus Mind Math Instruction Book Level – 2: Step by Step Guide to Excel at Mind Math with Soroban, a Japanese Abacus
ISBN-13: 978-1941589038

Abacus Mind Math Level – 2 Workbook 1 of 2: Excel at Mind Math with Soroban, a Japanese Abacus
ISBN-13: 978-1941589045

Abacus Mind Math Level – 2 Workbook 2 of 2: Excel at Mind Math with Soroban, a Japanese Abacus
ISBN-13: 978-1941589052

LEVEL – 3

Abacus Mind Math Instruction Book Level – 3: Step by Step Guide to Excel at Mind Math with Soroban, a Japanese Abacus
ISBN-13: 9781941589069

Abacus Mind Math Level – 3 Workbook 1 of 2: Excel at Mind Math with Soroban, a Japanese Abacus
ISBN-13: 9781941589076

Abacus Mind Math Level – 3 Workbook 2 of 2: Excel at Mind Math with Soroban, a Japanese Abacus
ISBN-13: 9781941589083